D1266478

TO
INFINITY
AND
BEYOND

Also by Neil deGrasse Tyson

Starry Messenger

Cosmic Queries

A Brief Welcome to the Universe

Astrophysics for Young People in a Hurry

Letters From an Astrophysicist

StarTalk Young Readers Edition

Accessory to War

Astrophysics for People in a Hurry

Welcome to the Universe

StarTalk

Death by Black Hole, and Other Cosmic Quandaries

The Pluto Files

Origins

The Sky Is Not the Limit

TO
INFINITY
AND
BEYOND

A Journey of Cosmic Discovery

NEIL DeGRASSE TYSON
LINDSEY NYX WALKER

Washington, D.C.

Since 1888, the National Geographic Society has funded more than 14,000 research, conservation, education, and storytelling projects around the world. National Geographic Partners distributes a portion of the funds it receives from your purchase to National Geographic Society to support programs including the conservation of animals and their habitats.

Get closer to National Geographic Explorers and photographers, and connect with our global community. Join us today at nationalgeographic.org /joinus

National Geographic Partners, LLC
1145 17th Street NW
Washington, DC 20036-4688 USA

For rights or permissions inquiries, please contact National Geographic Books Subsidiary Rights: bookrights@natgeo.com

Copyright © 2023 Curved Light Productions, LLC. All rights reserved. Reproduction of the whole or any part of the contents without written permission from the publisher is prohibited.

Some sections adapted from essays in *Natural History* magazine: Part One's "Cosmic Conundrum: The Coriolis Force" and Part Two's "Jupiter" adapted from "The Coriolis Force" (March 1995); Part Two's "The Tidal Force" adapted from "Tides and Time" (November 1995); Part Three's "Shocking Truths" adapted from "Shocking Truths: If You Break the Sound Barrier, You Can Make Quite a Stir" (September 2006).

NATIONAL GEOGRAPHIC and Yellow Border Design are trademarks of the National Geographic Society, used under license.

ISBN: 978-1-4262-2330-3

Printed in China

23/RRDH/1

To the explorers among our species and anyone else
with the audacity to pursue that which scares them

CONTENTS

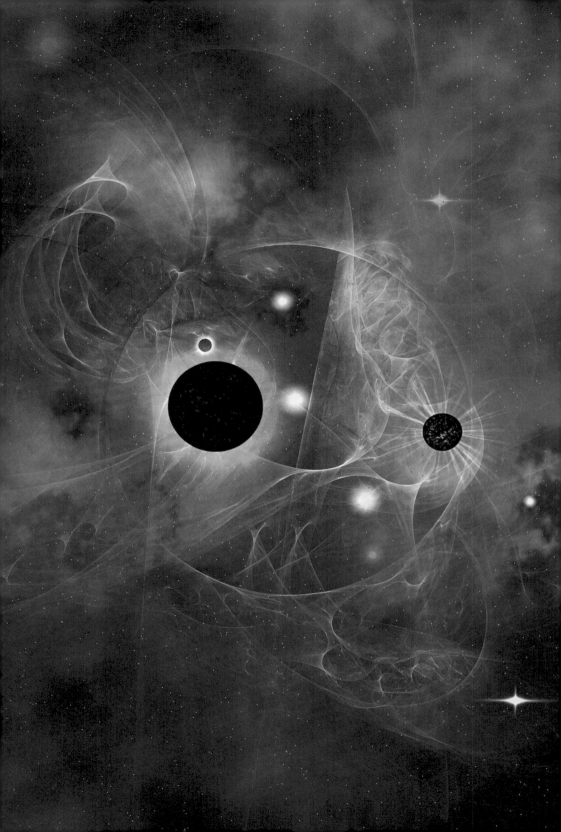

INTRODUCTION

THE COSMIC ODYSSEY

O nce upon a time, before humans understood what lay above the clouds, the realm of sky and stars was inhabited only by gods and explained only by myths and fables. But a series of discoveries, intermingled with fits and starts, wrong ways and dead ends, ultimately shattered those primal beliefs, empowering humankind with the knowledge to unveil strange and humbling truths. The odyssey of cosmic discovery had begun—and it has continued on ever since. Slowly but surely, a new universe emerged—one roiling with molecules, lurking with monstrous black holes, snaking with voids and galaxies of every size and shape, and hinting at untold mysteries yet to explore.

In this book, we invite you to embark on this journey with us—a journey of mind and body from Earth to infinity and

A vivid conceptualization of an energetic universe filled
with black holes, stars, and nebulae

PAGE 3: An enhanced view of the Milky Way galaxy as it might appear
from a spaceship traveling faster than light

PAGE 6: Baja California Sur and the Gulf of California as seen through a window
on the SpaceX Dragon Freedom crew ship, 261 miles above the Pacific Ocean

beyond. What empowered humans to escape our home, physically and intellectually, and soar into the unknown? What insights, what courage, what iconoclastic ideas, what technological failures and successes, carried us to the knowledge we have today? And what mind-blowing realizations at the edge of our understanding provide glimpses of a vast cosmos yet to be explored? It's a complex tale of people and planets, stars and spacecraft, a saga whose intricacies and eccentricities we explore in these pages.

Vastness, emptiness, darkness, coldness: grand and strange concepts ill fathomed by a comfortably warm-bodied, recently evolved, carbonaceous creature in a suburban solar neighborhood of the Milky Way galaxy. If you did not already know that Earth orbits the Sun, and not the other way around, you would have a hard time discovering that truth for yourself. If you did not know our solar system includes eight planets, hundreds of thousands of asteroids, and millions of comets, you might understandably assume that only Earth and the five planets visible to the unaided eye make up our little corner of the cosmos. To reach these milestones of knowledge, we had to leave the nest.

The force that keeps Earth whole, that tethers our Moon to Earth and Earth to our Sun, has also kept humans stuck beneath the clouds for nearly all our existence. We cannot easily escape Earth's gravity, which may be why the Wright brothers' first powered flight in 1903 and the Apollo 11 Moon landing in 1969 appear near the top of everyone's list of the greatest human achievements. Since then, thousands of satellites, hundreds of space probes, dozens of rovers, and even a helicopter have successfully launched from Earth, turning our eight-planet solar system into an explorer's backyard.

And that backyard continues to expand. In 2012, the Voyager 1 space probe went interstellar, escaping our solar system entirely—but not before transforming the mysterious planets and their moons into worlds of wonder. Voyager's final mission is not done, and it may very well carry on for longer than humanity endures. The tiny craft carries a golden record: an audio recording of songs and sounds of Earth and its species, bidding hello to anyone or anything that might intercept our greeting and serving as a collective plea to save us from our isolation. To acknowledge that we, small and feeble creatures, have a place within this great and ever expanding universe, Voyager carries forward the unending quest passed down from the first humans who wondered what more awaits our outstretched hands and upturned eyes.

Beginning in 2022, the James Webb Space Telescope continues the odyssey ever outward, sending us images that reveal the most ancient light ever seen and reminding us just how expansive our universe really is. Webb's First Deep Field unveiled thousands of faint and distant galaxies, including several that formed 13.7 billion years ago, images that bring us as close to the Big Bang as we've ever been able to reach. Try to imagine explaining that image to Newton or Galileo, whose radical new understanding of a geocentric universe upended all Christendom and shook the worlds of knowledge and belief. Imagine telling them that we are but one of trillions of planets in a universe with no tangible end. Imagine sharing with them how quantum physics and general relativity hint that not one, but countless, universes may exist beyond our own.

These are the stories that will unfurl as you read this book: a gravity-defying trajectory away from Earth through our solar

> The tiny craft carries a golden record: an audio recording of songs and sounds of Earth and its species, bidding hello to anyone or anything that might intercept our greeting and serving as a collective plea to save us from our isolation.

Taken by NASA's James Webb Space Telescope, this near-infrared image of galaxy cluster SMACS 0723 utilizes a natural effect called gravitational lensing.

neighborhood, out into the galaxy, and even farther. Along the way, we'll witness bewildering discoveries and unexpected conundrums that have awed and baffled the greatest minds throughout history, forcing them to rethink assumptions and

renovate worldviews. We may even come upon some concepts, out there in the abyss of the unknown, that inspire your own revolution of understanding.

Welcome to the cosmic odyssey—a thrilling, humbling, and thoroughly entertaining journey of discovery through space-time to infinity and beyond.

PART 1

LEAVING EARTH

"I know that I am mortal by nature, and ephemeral; but when I trace at my pleasure the windings to and fro of the heavenly bodies I no longer touch the earth with my feet: I stand in the presence of Zeus himself and take my fill of ambrosia."
—Ptolemy, *Almagest*

S tarry nights have guided, enlightened, and uplifted humanity for at least as long as people have been looking up. There's no telling who first dreamed of leaving Earth to explore what lay beyond, or who first wondered if a beyond did indeed lie anywhere at all. But we know that the allure of the Sun, the Moon, and the glistening dome of lights visible to our unaided eyes echoes across millennia of human culture.

The evidence is copious. Cave paintings and rock carvings that date back 40,000 years or more capture not only animals and hunters but also comets, meteors, and constellations in enough detail to track Earth's slow wobble on its axis: the precession of the equinoxes. In the 4,000-year-old *Epic of Gilgamesh* from ancient Mesopotamia (now Iraq)—a tale rife with adventures, heroes, villains, romances, and battles—constellations come alive as characters. In this, one of the oldest surviving works of literature, cosmic threads tie together the mortal and immortal realms, while time and distance are measured by the motions of the stars.

For thousands of years, humans reasonably assumed the Moon to be a flat disk of light that waxed and waned—until the 17th century, when Galileo Galilei dared turn his freshly

The Milky Way over Murtega Creek in Noudar Park,
Alqueva Dark Sky Reserve, Portugal

PREVIOUS PAGES: A striking visualization of Earth created with data layers from NASA's Blue Marble Next Generation image collection perfected telescope skyward, revealing a textured sphere with jagged mountains drenched in sunlight and sloping valleys cloaked in shadow. From that moment onward, the heavens and all the celestial objects therein became worlds—destinations—surfaces upon which humans might wander if only we could somehow cross the depths of space. Since then, but especially in the 20th century, scientists, engineers, daredevils, and politicians raced to new heights. Curiosity, competition, and war-hastened innovations carried us through the portal of our transparent atmosphere and beyond.

But before we could pierce the sky, before we could know it was navigable, humanity first had to discover what the sky is, what it is not, and where it ends—if at all. Our atmosphere, we would soon learn, was but a blue bubble that dissipates into an inhospitable vacuum devoid of the particles, pressure, and photons found in our comfortable earthly cocoon. Step by step, question by question, and discovery by discovery, we made our way through that bubble to whatever lay beyond.

In this section, we will track that odyssey, as humans ascended from the surface of Earth first with balloons, then airplanes, then jets, and finally with rockets all the way to the Moon. We will learn to navigate the atmosphere and overcome the force of gravity that for so long held the human imagination in its grasp.

From the dried and faded cave paintings of long ago to the freshly inked quantum equations of today emerges a story without end marked by cycles of curiosity, discovery, upheavals, and unlearnings as we ascend from one worldview to the next.

The cosmic journey begins.

EARTH'S ATMOSPHERE

Driven by our desire to soar like birds, we first took to the skies in story form. Tales of human flight proliferated through ancient legends and myths across many cultures. Alexander the Great is often depicted aloft on a winged chariot, pulled by four mythical griffins—creatures with a lion's body and eagle's wings. A Cowichan First Nation legend recounts the story of two boys who sang to their canoe to make it fly from the mountaintop where it was built, over their village, and out to sea. The ancient Sanskrit epic of *Ramayana* and other ancient Indian texts describe flying "vimanas," or self-propelled chariots of the gods.

But one of the most famous stories of human flight is portrayed in the celebrated Greek myth of Daedalus and his son Icarus. If anyone has ever cautioned you not to fly too close to the Sun, they were harking back to this ancient legend to warn you against the temptation of excessive risk and thrill.

To escape the island of Crete, Daedalus, a superb craftsman and aeronaut, fashioned two sets of wings from feathers bound with wax: one set for himself and one for Icarus. He urged his son to fly neither too close to the sea lest the dampness muddle the feathers, nor too close to the Sun lest the warmth melt the wax. But in his youthful rebelliousness, Icarus soared higher and higher toward the Sun until the binding wax of his wings melted; he plummeted to his death into the unforgiving Aegean Sea below. This cautionary tale predates our understanding of thermodynamics, aerodynamics, and atmospheric physics by two millennia, so we can forgive its inaccuracies. Icarus would surely have died, but the Sun's heat would not have been the cause.

We now know, as Icarus and the ancient Greeks who told his story did not, that Earth's atmosphere divides into five distinct layers: The troposphere, where all plants and animals live and breathe, packs three-quarters of all the air molecules and 99 percent of all Earth's water vapor. Nearly all weather occurs within this most dense layer—which reaches four to 12 miles above Earth's surface depending on latitude and season. And it continues to rise nearly 200 feet a decade due to Earth's climbing

"The Fall of Icarus" (1606–07) by Carlo Saraceni

temperatures. At the base of the troposphere, sea level, the average global temperature hovers around 59°F nowadays. At the top where it meets the stratosphere, however, the average temperature drops to minus 70°F and even chillier.

But if sunlight must first cross through the atmosphere before it reaches Earth, and if, by ascending, you are moving closer to the Sun, shouldn't that region be warmer than at sea level, as the author of Icarus's tale presumed? Anyone who has ever visited high elevations could answer that question. In fact, their experience might suggest an opposite scenario. Mountaineers know to calculate an average temperature drop of about 3.5°F for every thousand feet of ascent. From the base camp of Mount Everest, with a balmy average daytime temperature of 60°F in springtime, Sherpas and climbers can expect a dip to about minus 20°F by the time they reach the summit, if they survive at all. Turns out, temperature fluctuations felt here on Earth's surface have nothing at all to do with its distance from the Sun—something else entirely must be going on.

Let's first ask how much closer Icarus got to the Sun. If we assume he ascended 10 miles—higher than any airplane—with the Sun at a distance of 93 million miles from Earth, Icarus flew 0.00000001 percent closer to the Sun. Hardly enough to account for the story line.

As for the air temperature itself, we must first grasp the connection between light and heat. Our Sun emits light at every wavelength of the electromagnetic spectrum, powering nearly all life on Earth. The spectrum is endless, ranging from long-wavelength, gentle radio waves to short-wavelength, intense gamma rays. Between the two lies a narrow band we call visible light: the only part of the spectrum the human eye can perceive. Stars such as the Sun emit nearly half their energy as visible light, a small portion as ultraviolet (slightly shorter waves), and most of the rest as infrared (slightly longer waves), which we feel as heat. It's no coincidence, then, that most animals on Earth evolved sophisticated organs capable of perceiving these select portions of the unbounded electromagnetic spectrum.

Based on how Earth dwellers evolved, we might presume that extraterrestrials from another planet that circles a different type of star would develop perceptual organs exquisitely tuned for that portfolio of light. Aliens residing on planets circling cool, small, red dwarf stars—the most common type of star in our galaxy—may see their world awash in infrared, and struggle to perceive the higher-frequency color blue.

Temperature is simply a measure of molecular vibrations, and all molecules vibrate. (This is an important fact to pocket as we continue on our cosmic journey.) Anything with a temperature above absolute zero—that is, everything in the universe, including frigid icebergs and matter in the darkest, deepest expanses of space—emits electromagnetic energy. At higher temperatures, the emitted mix of electromagnetic energy favors the shorter wavelengths of light. Lower temperatures favor longer wavelengths. The Sun, with its average surface temperature of about 10,000°F, peaks in the visible light frequencies, as would anything else of a similar temperature. Meanwhile, nearly everything on Earth, including our 98.6°F bodies and the planet's surface itself, radiates primarily in the longer (and invisible) infrared frequency. That's why your evening campfire can still feel warm the next day even though it no longer visibly glows. As the once red-hot lump cools, the peak frequency of the light it emits shifts down to longer wavelengths and eventually exits the visible spectrum. The chunk of coal remains toasty warm, suffused with infrared energy long after you've stuffed yourself with s'mores.

Have you noticed that the warmest time of day is several hours after noon, rather than at noon itself, when the Sun is at its highest in the sky? The same infrared light we feel but can't see in campfire coals is what causes a sweltering summer afternoon. The atmosphere absorbs some infrared light from the Sun and transmits the rest to Earth's surface. But the largest assault of sunlight comes from visible light, which penetrates our atmosphere unscathed. The simple and profound fact that our atmosphere is transparent to visible light is what allows us to view the Sun, the Moon, and the constellations at all.

Small amounts of the Sun's shorter-wavelength UV light also penetrate the atmosphere—and cloud cover, which is why melanin-challenged people smear themselves with sunscreen, even on gloomy days, to avoid sunburn and ultimately skin cancer. When we turn our faces to the Sun, the visible and UV light collide with the molecules in our skin, exciting electrons within it, converting that motion into heat, and emitting that heat as infrared radiation. Similarly, once they absorb the various wavelengths of radiation, molecules on Earth's surface are transformed into infrared and are reemitted by the ground. That infrared energy then radiates back up through the atmosphere, warming the infrared-absorbent air. A July day feels hot not because the Sun heated the air from above, but because the ground heated the air from below. For this reason, the warmest part of the troposphere sits just above Earth's surface.

On its journey back into space, most of the infrared emitted by Earth collides with, and is absorbed by, select molecules in the atmosphere whose bonds cause them to bend and vibrate when they encounter infrared. Having absorbed the energy, these molecules then reemit it in all directions, including back toward Earth, where it is reabsorbed and reemitted. We call this continual back-and-forth cycle the greenhouse effect. It's the same thing, on a smaller scale, that happens within actual greenhouses and all cars with closed windows. Visible sunlight penetrates the transparent glass and is converted to infrared within, which then is blocked from escaping by the very same windows that allowed the visible light to penetrate. This raises the temperature of the interior air well above that of the external air, creating a contained microclimate comfortable for tropical flowers and deadly for unattended pets and children. Unless you're in the habit of transporting hibiscus plants or fiddle-leaf figs around town, you would be wise to keep your car windows cracked on even a mildly sunny day.

Fact is, life on Earth benefits from a mild greenhouse effect. Without it, Earth's average temperature would remain below freezing, and our planet's surface would evolve into an icy tundra with no life as we know it. Fortunately for us, our atmosphere largely stabilizes the day-to-night temperature swings via air currents. On the Moon, where there is no atmosphere, the surface

> Fact is, life on Earth benefits from a mild greenhouse effect. Without it, Earth's average temperature would remain below freezing, and our planet's surface would evolve into an icy tundra with no life as we know it.

temperature swings wildly from a scorching 250°F during the day to a frigid minus 200°F at night.

So, what would have actually happened to poor Icarus? His first folly was of aerodynamics: He could never have achieved liftoff to escape Crete in the first place. We can take one cue from cherubs and condors. They both sport wings and they weigh about the same as each other. But condors actually fly, without the help of Renaissance painters. Therefore, if a cherub were to fly, it would need the 10-foot wingspan of a condor. Scaling up to the weight of an adult human, it would follow that Icarus would need wings 10 times larger still—and suitable chest muscles to flap them. So with any attempt to fly, he would have just fallen flat on his face.

Holding that aside, had Icarus indeed flown closer to the Sun, his body and wings, far from melting, would have frozen on ascent and thereby doomed him to that same fatal tumble. In 1920, famed astrophysicist Sir Arthur Eddington offered a more charitable interpretation of this legend: "Perhaps there is something to be said for Icarus . . . I prefer to think of him as the man who certainly brought to light a constructional defect in the flying-machines of his day."

BEYOND THE TROPOSPHERE

The troposphere, from the Greek word *tropo,* meaning "change" or "turn," is characterized not only by the changeable weather found within it but also, notably, by its tendency to lose heat with altitude. The next layer of our atmosphere, the stratosphere, is characterized by the exact opposite thermal phenomenon: As stratospheric altitude increases, temperature increases. Something must be absorbing energy there, increasing the vibration rate of air molecules. The culprit? The stratosphere is home to the ozone layer—the region of highly concentrated

The Moon rises beyond the orange-colored troposphere—the lowest and most dense portion of Earth's atmosphere—which ends at the tropopause, the boundary between the orange- and blue-colored atmosphere.

three-atom oxygen molecules (O_3) that absorb nearly all the Sun's most harmful ultraviolet rays. An ultraviolet photon carries the precise energy required to break apart an ozone molecule, turning O_3 into O_2 plus O. Curiously, the same ultraviolet light will then break apart the O_2, leaving O plus O, allowing each to recombine with stray O_2 molecules to recover the lost ozone:

$$O_3 + UV \rightarrow O_2 + O$$
$$O_2 + UV \rightarrow O + O$$
$$O + O_2 \rightarrow O_3$$

In other words, the ozone layer's turbulent dance of tearing apart and re-forming molecules keeps it in equilibrium with the Sun's flux of ultraviolet (UV) light. Absent this protective layer, the Sun's ultraviolet energy would do untold damage to the DNA of anything alive on Earth's surface.

The next highest layer, the mesosphere, is where meteors burn up in the dazzling displays we call shooting stars. Above that is the thermosphere, through which the International Space Station (ISS) and thousands of satellites orbit Earth. While the thermosphere is one-millionth the density of sea-level air, it hosts the most solar activity. This layer is home to the beautiful northern and southern lights: the aurora borealis and aurora australis.

The thermosphere is so named—*thermo-* from Greek for "hot"—because, in one sense, it's the hottest layer of them all. We measure temperature from molecular vibrations, and we measure heat by adding up the vibrational energy from all the resident molecules. Molecules vibrate most rapidly in this upper layer. But those molecules are so sparse that they would barely register on a human body. In fact, if you visit the thermosphere without first donning a space suit, you'll likely suffocate from lack of oxygen before feeling any significant warmth.

Past the outskirts of the thermosphere lies the exosphere, our atmosphere's final, outer layer, reaching far beyond all the other regions combined and consisting of only trace amounts of atmospheric molecules.

THE WEIGHT OF AIR

Have you ever used the phrase "light as air"? If so, you were probably describing something without realizing just how hefty air is. Yes, air has weight. That weight is what we call air pressure. Obvious next question: What is pressure?

All day, every day, pressure shows up in your life in both important and unimportant ways. How sharp are your kitchen knives? How comfortable is the chair on which you sit? Why do your high heels hurt so much more than flat shoes? Pressure is simply force (the weight of any object such as your knife, your butt on the chair, your whole body) divided by the area (the edge of the knife blade, the chair cushion as it contours your butt cheeks, your poor assaulted toes) over which the force is exerted. In other words, the smaller the area, the higher the pressure exerted by a given force.

So, if you must cross a frozen lake, how might you do it? By putting this principle into action. People with tiny feet are much likelier to break through a thin layer of ice than people who weigh the same but wear snowshoes, spreading out their weight across a larger area. Still worried? Your very best bet would be to lie down on your belly and inch your way across, spreading your weight across the area of your entire body.

If the weight of air results in air pressure, that means air exerts a measurable force. Imagine you're at sea level—say, your local beach—and you've brought an empty glass column one inch by one inch square. Place one end on the ground and let the top end magically grow straight up like Jack's beanstalk until it reaches the outer limit of Earth's atmosphere. You have now excised, cookie-cutter style, a long gaseous pillar. If you took all the air in that column and put it on a scale, it would weigh nearly 15 pounds. That's 15 pounds pushing against every square inch

A polar bear intuits the physics required to move over pack ice near the Arctic National Wildlife Refuge in North Slope, Alaska.

of your body all day, every day. But if someone placed 15-pound weights on every square inch of your body, you wouldn't be able to lift your chest to breathe. What's going on here? How can humans survive under such immense pressure?

Pressure in a fluid manifests in every direction, not just downward. Strange as it may seem, air is classified as a fluid. By definition, a fluid takes the shape of its container, which both gases and liquids clearly do. So, the air pushing down on your body must also manifest in every direction, like any other fluid. The pressure pushing down is the same as the pressure pushing up, down, sideways, and all around in all directions. As a result, all those forces are neutralized, and, presuming you have air inside your lungs, you don't feel the weight of the air at all.

What if you somehow removed the counterbalancing forces? Anybody who's ever used a suction cup has performed that experiment. When you plunge a suction cup onto a hard, smooth, flat surface, you'll discover it's difficult to pull it back up. In fact, that's precisely the purpose of a suction cup. The larger the cup, the harder it is to release it from the surface, even though no glue or other sticky substance keeps it there.

Why? Because suction is not a force; it's the response of atmospheric pressure to a vacuum. Fifteen pounds per square inch of air pressure is now pushing the rubber suction cup flush to its affixed surface, creating a vacuum and removing any counterbalancing force behind it. So, if your suction cup is 10 square inches, you're contending with 10 cookie-cutter columns of atmospheric pressure. Ten times 15 pounds means you must now exert 150 pounds of force to unstick the cup, or about the weight of an adult human.

Because air pressure acts equally in all directions, your suction cup works in any orientation. In many heist films, a sneaky burglar will use suction shoes and gloves to scale walls and ceilings to evade detection or avoid tripping a laser alarm. Ideally, some attached mechanism would pump air into and out of the suction shoes to save our protagonist from the exertion of unsticking 150 pounds of pressure with every step.

But fluids do more than take the shape of their containers. Another of their notable properties is the buoyant force within them. Around 250 B.C.E., the Greek mathematician Archimedes of Syracuse reportedly exclaimed, "Eureka! (I have found it!)" after discovering buoyancy while passing a pleasant hour in one of ancient Greece's public baths.

According to the ancient tale, the king of Syracuse hired a goldsmith to craft a crown from a lump of gold that the king had weighed beforehand. The goldsmith accepted and soon executed the task. The king, suspicious and greedy, wanted to be certain the goldsmith hadn't stolen some of the gold and replaced it with less valuable silver. So he turned to his cousin Archimedes, who was renowned for his achievements in mathematics, to devise a strategy that would ascertain the crown's authenticity and the goldsmith's integrity. Where better to ponder a conundrum than in the bath? Upon lowering his body into a full basin of water, Archimedes noticed that the tub overflowed, whereupon he realized that the volume of overflowed water would equal the volume of his submerged body parts.

Because Archimedes knew the gold's original weight, and he now knew how to measure the volume of an irregularly shaped object, he could determine authenticity by comparing densities. Density is the mass of an object (or its weight in this instance) divided by its volume (or size). Archimedes obtained a lump of pure silver and a lump of pure gold, each the same mass (or weight) as the crown. He then submerged the lump of gold in a bowl filled with water and measured how much water the lump displaced. He then repeated the experiment with the silver lump and compared the two measurements. Because silver is less dense than gold, the silver lump was larger and displaced more water than the gold lump of the same mass. To complete his experiment, Archimedes finally lowered the king's crown into the same water-filled bowl. Had the crown been constructed of pure gold, it would have displaced the same amount of water as the unformed lump of pure gold. But it didn't: It displaced more. Thus, using his clever new method, Archimedes ascertained that the goldsmith had indeed cheated the king—or so the story goes.

This method works for any object in any fluid. (Good to know in case you're ever called to verify the authenticity of a golden crown.) But there's even more going on here. In his work *On Floating Bodies,* Archimedes wrote that any body wholly or partially immersed in a fluid experiences an upward force equal to the weight of the fluid displaced—the cause of buoyancy. Anything will float if it weighs less than the total weight of the liquid it displaces.

Exploiting the buoyant force revolutionized global industry, politics, and society. For example, a lump of steel will sink as readily as a lump of wood will float; we know this intuitively. And yet, since the mid-1800s, wooden warships have been upgraded to ones crafted of steel and iron, laden with weaponry and fleets of sailors. A ship once easily sunk by fire or cannon became resilient to attack. A century later, sea crossings became a glamorous leisure rather than a treacherous voyage. Millions of people happily embark on thousand-foot-long cruise ships made of welded steel every year. These enormous vessels float because their total volume—including all the air of its hollowed innards—weighs less than the weight of the water they displace.

What about the humans aboard those ships? On their own, our muscles and bones would sink, whereas our fat would float. But human adults are about 60 percent H_2O (and human infants about 80 percent). So, a bodybuilder thrown overboard will more readily sink, while your average person will more easily float.

Because a human body's overall density is similar to that of water, the volume of water you displace weighs about the same as you do, a fact that makes you practically weightless in water—not bobbing high like Styrofoam or cork, but not sinking like a rock, either. But in the Dead Sea, where the water is almost 10 times saltier than the ocean, everybody floats—even your fitness trainer with the six-pack abs. The salt creates a much denser medium than ordinary seawater, resulting in a much stronger buoyant force acting on your body.

Since Archimedes' principle applies to any fluids, the buoyant force operates in both ocean and air. Let's return to the one-square-inch column that you've cookie-cut through the atmosphere. At the bottom of that column sits 15 pounds of pressure. As you ascend, however, less and less air presses down on you. Thus, the air pressure drops.

In the Dead Sea, where the water is almost 10 times saltier than the ocean, everybody floats—even your fitness trainer with the six-pack abs.

In 1644 Evangelista Torricelli proposed a revolutionary claim: "We live submerged at the bottom of an ocean of the element air, which by unquestioned experiments is known to have weight." He re-created an experiment that had stumped even the great Galileo. If you fill a 30-foot-long tube with water, and then invert the tube into a basin of more water, the tube will empty only a little of its contents into the basin, leaving an empty space at what is now the top of the tube. Galileo had argued that a vacuum in the top of the tube somehow tugged on the water to prevent it from emptying into the basin, but he failed to prove his hypothesis.

In response, Torricelli reasoned that the vacant space at the top of the tube must indeed be a vacuum, but that its vacuum

status was irrelevant to what was going on. Rather, he asserted, the air all around and above was pushing down on the exposed surface of the water in the basin. In turn, the water in the basin exerted upward pressure on the water in the tube, preventing it from emptying completely. Torricelli ultimately perfected this same experiment with mercury instead of water. Mercury is nearly 14 times denser than water, allowing for a much smaller tube and basin.

The higher the atmospheric pressure on the pool of mercury in the basin, the higher up the tube the mercury creeps. The lower the atmospheric pressure, the more mercury spills out of the tube into the basin. Mark the height of the tube in inches to quantify what's going on, and behold: the world's first mercury barometer. (So the next time you hear a weatherperson announce the barometric pressure in "inches of mercury," you'll know exactly what they're talking about. Credit Torricelli.)

Imagine how puzzled Galileo might have been over a simple demonstration with a drinking straw. When you dip the straw into a beverage and cover the top with a finger, most of the liquid remains inside when you withdraw the straw—not because of some mystical vacuum between your finger and the liquid within the straw, but because all the air pressure outside the straw pushes up from below. Your finger cuts off the pressure from above that would otherwise equalize the effect on the liquid.

Shortly after Torricelli's discovery, a French mathematician named Blaise Pascal, fascinated by this hypothesis of air pressure on the basin, proposed that if air was pushing down on the basin, then a place with less air, such as the top of a mountain, would push down less, allowing more mercury to spill out of the tube and into the basin. He convinced his brother-in-law to carry

A depiction of Italian mathematician and physicist Evangelista Torricelli's barometer experiment (1644), in which the existence of atmospheric pressure is experimentally demonstrated an enormous mercury barometer to the top of the tallest nearby mountain, Puy de Dôme, taking measurements along the way. To his delight, more and more mercury emptied into the basin as he climbed, showing that atmospheric pressure drops with the increase in altitude.

Emboldened with the discovery that air has a measurable weight, and that the weight lessens with altitude, Earthlings soon contrived new ways to buoy our way through the skies above, and soar like Icarus before his final moments.

DREAMS OF ASCENDING: THE BALLOONATICS

The design of the first vessels to carry humans upward through the atmosphere derived from ancient Chinese technology: Kongming lanterns, floating balloons that housed small oil lamps or candles. Dating back to about 80 C.E., these lights were originally used in war to send signals across the landscape to one's own army or to confuse one's opponents. Today they're best known as sky lanterns and pop up across the world in celebrations such as Diwali, a Hindu festival, and Yi Peng, a lunar festival in northern Thailand.

The original Kongming lanterns were cocoons of paper or cloth that trapped the heated air above a flame. When a substance is heated, its molecules vibrate faster (recall that temperature is simply a measure of this movement). The energized, rapidly moving air molecules require more space to wiggle and jiggle, and so the volume they occupy expands as they migrate outward away from one another, leaving the cavity within the balloon less dense than the surrounding air. Inside the lantern, the capsule of hot air rises within the surrounding ocean of denser air, its candle payload in tow. The lantern continues to

Golden lanterns illuminate the night sky during the
Yi Peng lunar festival in northern Thailand.

ascend until the ever thinning surrounding air reaches the same density as the entire lantern itself, or until the candle burns out.

If you're asked to recall the first aeronautical pioneers, the Wright brothers might come to mind. But a century and a half before Orville and Wilbur took flight in 1903 at Kitty Hawk, North Carolina, a pair of French brothers had already kick-started human aviation history with their "aerostatic globe," better known as a hot air balloon. These vehicles rely on the same buoyancy physics as the Kongming lanterns of ancient China, but they use a cocoon cavernous enough to haul a payload of humans dangling below.

In 1783 Joseph-Michel and Jacques-Étienne Montgolfier tested the first such balloon in southern France. They also recruited the first known aeronauts: a sheep, a duck, and a rooster. The peculiar trio returned safely to Earth after an eight-minute, two-mile journey, ushering in the era of human flight.

Almost a century later, meteorologist and astronomer James Glaisher teamed up with expert balloon pilot Henry Coxwell for a near-fatal aeronautics experiment. Glaisher was determined to understand just how high a balloon could carry a person and what could be learned about the atmosphere and air pressure along the way. In his 1871 book *Travels in the Air,* Glaisher asked, "Do not the waves of the aerial ocean contain, within their nameless shores, a thousand discoveries destined to be developed in the hands of chemists, meteorologists, and physicists?"

By Victorian times, experimenters recognized that a balloon pumped full of a gas that was natively lighter than air could ascend higher and more rapidly than plain hot air. And you wouldn't have to heat it. Most balloonists of this time used common coal gas—the stuff that powered their kitchen stoves—as it contained a low-density mixture of hydrogen, methane, and carbon monoxide. With their coal-gas balloon, Glaisher and Coxwell attained an altitude of more than 35,000 feet before losing consciousness from lack of oxygen in the low-pressure air as their skin blackened with frostbite. They also caught a case of decompression sickness, colloquially known as the bends—the same affliction scuba divers experience if they resurface too quickly. They did this for science. And though they neared death in their experiment, they fared much better than Icarus, who actually died trying.

According to Coxwell and Glaisher's barometers and the other instruments on board, they either reached the stratosphere or came very near it. Had they been able to survive a longer ascent, they would have discovered interesting data within the next layer of Earth's atmosphere. For every buoyant balloon, however, there's an altitude above which it ceases to ascend. As long as the balloon and its payload combined are less dense than the surrounding ocean of atmosphere, it rises. Arriving at the elevation of equal density, though, it will merely bob along like a buoy at the surface of the sea.

These days, meteorologists routinely launch weather balloons—minus candles, farm animals, or humans—but equipped instead with instruments designed to monitor pressure, temperature, and relative humidity. These balloons enable the hourly weather reports we now rely on to know if the baseball game might be rained out, if NASA will proceed with a rocket launch, and whether you should don a sun hat or earmuffs before you walk out the door.

> These balloons enable the hourly reports we rely on to know if the baseball game is rained out, if NASA will proceed with a rocket launch, and whether you should don a sun hat or earmuffs.

FELIX BAUMGARTNER AND THE EDGE OF SPACE

On October 14, 2012, a century and a half after Glaisher and Coxwell's treacherous experiment, a skydiving Austrian daredevil named Felix Baumgartner smashed several world records in a feat that headlines across the country extolled as the "jump from the edge of space." His ascent to the stratosphere in a helium balloon and free-fall return to Earth's surface set a new record for the highest human-piloted balloon ascent and highest skydive (from about 24 miles up); he became the first person to break the sound barrier without help from an engine. But he did not jump from space.

Baumgartner plummeted to Earth at 844 miles an hour, a feat made possible by the scarcity of air molecules in the stratosphere and upper troposphere that would have slowed him

Austrian pilot Felix Baumgartner readies himself to jump from
29,455 meters during the second crewed test flight for
Red Bull Stratos in Roswell, New Mexico, in 2012.

down. For comparison, the higher air density of the lower atmosphere limits the terminal velocity of a human body—the fastest
achievable skydiving speed—to about 120 miles an hour for a
full-belly fall. Baumgartner's altitude was indeed sufficiently
high that he had to wear a pressurized suit during his stunt. But
he's no astronaut—that is, he didn't come close to what we call
"space." To warrant that designation, he would have had to jump
from more than twice as high. So, the jump from "the edge of
space," sponsored by the highly caffeinated drink Red Bull, was
given a highly exaggerated tagline. Although he achieved an

incredible feat that captivated the world, if Earth were scaled down to a schoolroom globe, Baumgartner jumped from just a millimeter up.

According to most people who care about such things, "space" begins at 100 kilometers (62 miles) above sea level. This altitude, known as the Kármán line, is named after the Hungarian American aerodynamicist Theodore von Kármán, who was the first to define this boundary. In his autobiography (published posthumously, in 1967), he described it as "a physical boundary, where aerodynamics stops and astronautics begins." In simple translation: Where there's no air, an airplane no longer works because it relies on air passing over its wings to gain lift. Above the Kármán line, you need rockets.

Kármán took the definition one step further when he added, "Below this line, space belongs to each country. Above this level there would be free space." But as it turns out, Earth's atmospheric layers don't divide into rounded metric units. So, though conveniently rounded, the 100-kilometer "boundary" is actually quite fuzzy; Kármán himself had recommended a lower altitude. The matter has been debated for decades as new understandings of Earth's atmospheric profile have emerged.

THE BILLIONAIRE "SPACE RACE"

In July 2021, half a century after the first humans landed on the Moon, a couple of billionaires captured the world's attention in what was portrayed as a new space race. This time around, the race was not a war-fueled competition to discover what no human had seen before, but rather a rivalry among the world's 0.0001 percent to commercialize spaceflight for personal profit. Richard Branson, founder of the Virgin Group and the spacecraft company Virgin Galactic, became the first billionaire to fly to (one definition of) space. He and his crew on the Unity spaceplane reached an altitude of 54 miles (87 km), where they enjoyed a few minutes of weightlessness before gliding back to Earth. Afterward, on terra firma, Canadian celebrity astronaut Chris Hadfield pinned the coveted astronaut wings to Branson's chest, indicating membership in an extremely exclusive club. But is Branson really an astronaut?

Nine days later, Jeff Bezos, founder of Amazon.com and the spacecraft company Blue Origin, launched his own suborbital vehicle, New Shepard (named for Alan Shepard, the first American in space). It reached just above the 100-kilometer (62-mile) Kármán line; he, too, received a pair of astronaut wings.

Are these guys really space travelers? Is there a true, clear boundary between down here and out there? Where does space really begin?

According to the Fédération Aéronautique Internationale (FAI), the nongovernmental international regulatory body of air sports, outer space begins at the 100-kilometer Kármán line. The U.S. Federal Aviation Administration, the U.S. military, and NASA, however, currently place the edge of space lower, calling anybody who's soared higher than 50 miles (80 km) an astronaut.

Looks like the urge to use round numbers for the Kármán line is strong, whether in metric or imperial units. Kármán himself proclaimed the edge of space to be 275,000 feet high—about 52 miles up. Perhaps the United States' definition of space should have been named after Kármán.

Whatever the intent behind these definitions, the fact remains that none have practical, scientific implications. The boundary, whatever it is, remains fuzzy; there's just no good distinction between "space" and "not space." In fact, scientists recently detected wisps of Earth's atmosphere beyond the orbit of the Moon. So, if we want to define "not space" by "no atmosphere," then our Moon is still within Earth's atmosphere, and nobody has ever ventured far enough to become an astronaut.

Ultimately, both Bezos and Branson were stripped of their astronaut wings—not because of any issues about altitude but because the U.S. Federal Aviation Administration imposed a stricter definition of "astronaut" right after their flights, dealing a major blow to space billionaires everywhere. Now, an astronaut, in addition to reaching an altitude beyond 50 statute miles, must also have "demonstrated activities during flight that were essential to public safety, or contributed to human space flight safety."

FROM AIRCRAFT TO ROCKETS

If we wish to define the boundary of space as a region where regular airplanes can no longer remain aloft, as Kármán himself suggested, we should understand how airplanes accomplish their feat in the first place.

Airplanes can fly, not in spite of air density, but because of it. As a plane plows through the atmosphere, it crashes into

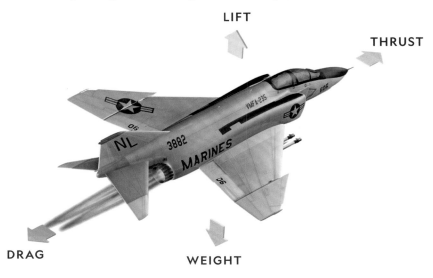

These arrows visualize the forces—lift, weight, thrust, and drag—that act on an airplane during flight.

countless molecules. Airplane wings are curved on top and flatter on the bottom, causing the air molecules to flow faster over the top of the wing than along its underside, which in turn creates lower pressure above than below. This follows Bernoulli's principle, named after Swiss mathematician Daniel Bernoulli, which posits that the faster a fluid moves, the less pressure it exerts.

Fluids tend to move from areas of high pressure to areas of low pressure, and air behaves like a fluid. When a lower-pressure environment exists above the wing, air particles below the wing rush upward to create lift. At the same time, gravity pulls the plane downward and friction from the air molecules causes drag. Not to worry, though. On the other side of the equation, huge engines—propellers or jets—generate thrust as they impel the plane forward, creating and maintaining the pressure difference on the wings and thus sustaining lift. Commercial airplanes usually fly above 30,000 feet, where there's the least amount of drag from atmospheric particles but just enough air density to keep the plane aloft in excess of 500 miles an hour without burning huge quantities of fuel.

An important difference between airplanes and rockets, by the way, is that airplane engines burn their fuel with the help of oxygen drawn from the atmosphere. Rockets that fly beyond Earth's atmosphere must use self-contained oxidizers. Rocket engines can burn fuel all by themselves, without help from atmospheric oxygen at all.

ROCKET SCIENCE AND MAX Q

To a rocket scientist, Earth's atmosphere is merely an obstacle standing between an astronaut and outer space. Beyond the troposphere, the air continues to grow thinner and thinner, and aircraft experience less and less lift. But if the plane has an extremely powerful engine, it can generate enough thrust to

WHY DO BASEBALLS CURVE?

Whether or not they know it, baseball pitchers exploit the same principles that keep airplanes aloft. A pitcher's well-executed curveball befuddles a batter by creating a pressure difference from one side of the spinning ball to the other. Just as with an airplane's wing, the air on the higher-pressure side of the ball pushes the ball toward the side with lower pressure. Perfecting a curveball requires diligence. But with enough practice, the incoming pitch should appear to be a typical, down-the-pipe fastball before its trajectory curves away from where it might have otherwise arrived, leaving even the most seasoned sluggers on their toes or with a strike on their count.

An effective yet illegal way to improve one's curveball would be to scuff up the ball a bit on one side, so that it "throws" more air than the stitching alone can provide. Another factor is the air density through which the ball travels. In places such as Denver's mile-high Coors Field, the air is 20 percent thinner than at sea level, rendering curveballs less effective than down at, say, Boston's Fenway Park. And in the rarefied Martian atmosphere, at around one percent of Earth's pressure, an attempted curveball wouldn't fool any batters at all.

Everything described that applies to baseballs also applies to soccer balls. When a skilled kicker boots a ball with a technique to intentionally add spin, remarkable curved trajectories ensue, sometimes fooling an entire defense to stand in the wrong place.

counteract gravity on its own, without the help of Bernoulli. And that is exactly what rockets do.

During a rocket launch, at about one minute after liftoff, the flight deck announces the moment of "max q"—code for maximum dynamic pressure. This marks the most dangerous moment

continued on page 46

MARS INGENUITY

Airplanes and helicopters exploit the same forces to stay aloft. A helicopter's rotating propellers, or rotor blades, provide both lift and thrust. An airplane must first barrel down a long runway to move enough air over its wings to put Bernoulli's principle into effect. A commercial jet will near 200 miles an hour before enough lift carries its passengers upward. A helicopter, on the other hand, whips its long blades around its otherwise stationary cockpit. Like airplane wings, helicopter blades are typically curved on top and flatter on the bottom, designed to create a pressure difference that provides lift. They can also be pitched at an angle to achieve the same effect. As long as a helicopter's blades provide more lift than the combined weight of its body and passengers in tow, it will stay aloft and can land or take off from nearly anywhere without ever needing a runway. Helicopters trade speed and power for convenience and maneuverability.

NASA's Perseverance rover on Mars carried on board a teensy, four-pound helicopter aptly named Ingenuity (shown at right in an artist's rendering). In the spring of 2021, Ingenuity became the first aircraft flown on another planet. Because Mars's atmosphere is a hundred times thinner than Earth's, that flight was equivalent to flying 16 miles above Earth (nearly three times higher than a jumbo jet). For the record, the highest recorded helicopter flight was about 5.5 miles up, to the top of Mount Everest, where the air is merely a third the density at sea level; beyond that, in thin air, the burden of gravity overcomes the thrust achievable from the rotating blades. So, Ingenuity had to be a featherweight and have proportionately enormous blades that could rotate up to 10 times faster than your usual Earth helicopter.

continued from page 43

of the launch. Atmospheric drag creates stress on a rocket and increases with the rocket's speed. As a rocket punches through dense atmosphere, the atmosphere punches back (and as we know by now, that punch has weight behind it). If the rocket moves too fast through this region, the vehicle could break to bits. As the rocket's altitude increases, however, there's less and less air available to cause drag.

Immediately after a rocket launches, it moves slowly through thick, sea-level atmosphere. There's not much stress here, as the rocket hasn't picked up enough speed yet to cause concern. Several minutes later, it zips through hardly any atmosphere at all; not much stress here either. In between these two regions, there is a spot where the speed of the rocket and the air through which it plows create maximum stress. That point is max q.

LAUNCH LOCATIONS: THE RHYME AND REASON

When seeking orbit, the ultimate goal of a launched rocket is not up, up, and away (although it does appear that way to the average onlooker at ground level). You might assume NASA simply pointed a Saturn V rocket at the Moon in the late 1960s and sent it off in a straight line. In fact, the Apollo astronauts launched eastward from Cape Canaveral, Florida, and made one and a half orbits around Earth before setting out toward the Moon. Among the many reasons to start this way is to exploit Earth's rotation, which offers a thousand-mile-an-hour running start for free at the Equator.

The most efficient way to get from zero to 17,000 miles an hour—the speed at which a spacecraft achieves Earth orbit—is to utilize the speed of Earth's rotation. At the Equator (0°N, 0°S),

MARTIAN DUST STORMS

n the 2015 blockbuster sci-fi film *The Martian*, starring Matt Damon and based on the book of the same title by Andy Weir, astronaut Mark Watney is stranded on Mars, left for dead by his crewmates following a huge dust storm. Mars is known to have monster storms that sometimes last for months on end and occasionally engulf the entire planet. Marooned and desperate, Watney is left with no choice but to "science the shit out of" his survival until a rescue ship can save him. The storm's dangerous winds had converged on his team's launch site, bashing their rocket with debris and threatening to tip over the mission's ascent vehicle. Presuming Watney dead, the captain decides to escape the planet with her remaining crew.

High drama, indeed. But in the actual universe, a Martian dust storm would feel like a gentle breeze—not nearly strong enough to knock over a person, let alone an entire spaceship. Hollywood had overlooked the dynamic pressure of the Martian atmosphere. Though the gusts of a Martian dust storm can reach gale speeds, at a mere one percent of Earth's atmospheric pressure, the only real threat to a Martian explorer would be reduced visibility and perhaps a clogged air filter. If *The Martian* had depicted a perfectly accurate Martian dust storm, Mark Watney and his crewmates might have fumbled around in a reddish haze, groping for the ladder to their ascent vehicle.

But a dust storm isn't without threat. The smothering blizzards of sand can last for weeks at a time, blocking nearly all sunlight. If one such storm occurred at the beginning of their journey instead of at the end, it might have meant catastrophe for the crew. Their habitat's air filtration systems would have eventually clogged with dust as all their solar-powered batteries slowly died.

CHALLENGER DISASTER: A TRAGEDY AT MAX Q

"Roger, go at throttle up," space shuttle *Challenger* commander Dick Scobee announced 70 seconds after liftoff on the frigid morning of January 28, 1986. The dangerous dynamic pressure of max q was now behind them, and it seemed safe to ramp up the engines from 65 percent power to full blast toward space. Nothing but a bit more atmosphere stood between the crew and their mission. But Scobee's words would be the mission's final communication to the flight controllers on the ground. A few seconds later, a cascade of failures would dismember the ship in a torrent of smoke and fire, catapult the crew compartment into the Atlantic Ocean, and kill all seven aboard.

A lengthy investigation, during which no space shuttles flew, revealed the culprits: a faulty seal, a failed safety check, freezing weather, and max q. The fully fueled, 4.5-million-pound *Challenger* space shuttle consisted of four main components: two solid rocket boosters, one giant external fuel tank, and the orbiter itself. The two solid rocket boosters relied on rubber O-rings to keep the fuel encased within the chambers. The frigid conditions, however, compromised the elasticity of one of the rubber rings; the seal then failed, allowing hot gas to escape. The stress of a strong crosswind at max q—that most dangerous point of maximum dynamic pressure in a rocket's journey—had jostled the craft enough for the burning propellant to burst forth as violently as a blowtorch, igniting the unspent fuel. Seventy-three seconds after launch, the shuttle disintegrated.

Nearly two decades had passed since the Moon landing. Space travel was considered safe and routine. Regulation had grown lax, and although the engineers had warned of O-ring failure in extreme cold, the launch took place. Two years and eight months later, after imposing rigorous new testing protocols, NASA resumed human spaceflights.

that's about 1,000 miles per hour due east. During our planet's 24-hour daily rotation, the Equator (with the widest circumference of our planet) has to travel more miles than any other latitude with smaller circumferences. At Cape Canaveral (28°N), that speed drops to about 915 miles an hour. New York City (41°N): 780 miles an hour. London (51°N): 650 miles an hour. Oslo (60°N): 520 miles an hour. Meanwhile, at the North Pole (90°N), Santa Claus simply pirouettes in place.

Were you one of those rambunctious kids who enjoyed the thrills and fears offered by your playground's merry-go-round? If so, you probably convinced your friends to spin the disk as fast as possible while you clung on for dear life. You would have noticed that the faster the platform spun around, the more you felt you were flying outward. To compensate, maybe you crouched near the center of the merry-go-round and held onto the bars with all your might. That feeling of flying outward from a rotating object, whether you're at the edge of a carousel in a park or on the surface of a planet in the solar system, is called centrifugal force—although it's not an authentic force at all. It's just your tendency to want to fly off on a tangent, which feels like a force to you.

At Earth's Equator, where the surface moves the fastest, the centrifugal force is strongest. So, why aren't Ecuadorians, Singaporeans, Galápagos giant tortoises, and other equatorial residents clinging to the ground with Velcro for fear of flying off? One word: *gravity*. However, that doesn't mean the centrifugal force is absent. It shows up in people's reduced weight. Everyone weighs a little bit less on the Equator than they would anywhere else in the world. If Santa

> Why aren't Ecuadorians, Singaporeans, Galápagos giant tortoises, and other equatorial residents clinging to the ground with Velcro for fear of flying off? One word: *gravity.*

weighs 400 pounds at the North Pole, he will weigh only 399 when he delivers presents to Ecuador. Not enough to notice. Barely enough to comment on.

To complete this scenario: If Earth suddenly ceased spinning, then everyone in the world who was not otherwise attached to Earth's surface would fall over and roll due east at their appointed speeds. Anyone south of New York and north of New Zealand would break land speed records with their careening bodies. Meanwhile Santa's workshop continues business as usual.

Let's try a thought experiment (a favorite exercise of Einstein's, by the way). If Earth started spinning faster and faster, the centrifugal force would keep increasing, just as it would on your merry-go-round. There's a speed at which it will become so great that it will counteract gravity entirely; these forces then cancel each other out, rendering you weightless and hovering above the ground. That speed is about 17,500 miles an hour. If Earth rotated at that speed, a day would last an hour and a half instead of 24 hours. We've seen speeds like that before. It's no coincidence that this is about the same speed required to achieve low Earth orbit (LEO).

Centrifugal force sends riders into the air on a chain swing ride in Sweden.

SPEED VERSUS ACCELERATION

I n a memorable scene from the 1986 film *Top Gun*, following a flight on a supersonic fighter jet, Maverick (Tom Cruise) bellows to his friend Goose (Anthony Edwards), "I feel the need—" and together they yell, "—the need for speed!" and they execute a perfect high five. What Maverick and Goose didn't understand, though, is that speed was mostly irrelevant to their delight.

Reminder: At this moment, everyone at the latitude of New York City (including both authors of this book and all the people in the cities of Barcelona, Rome, Istanbul, and Beijing) is moving due east at 780 miles an hour on Earth's spinning surface. And as Earth orbits the Sun, we fly through space at 18 miles a second.

Objects moving at a constant speed do not notice any motion unless and until the speed changes. This change is called acceleration, and it can be positive or negative, with the negative version more commonly referred to as deceleration. A change in direction while in motion is another version of acceleration. So when speed or direction changes, every object—your body, a bicycle, a rocket ship—feels it and responds.

When we quickly accelerate forward, our bodies are thrown backward into our seats. When we bank a turn, we lean in the opposite direction. When we quickly decelerate, we're thrown forward—and if we've forgotten to wear a seat belt, we'll continue flying until something like a windshield or a tree arrests our trajectory. Sports car dealerships might advertise the top speed for a car, but the more interesting information is the car's ability to go "from zero to 60" superfast. That's acceleration.

So, in fact, when Maverick and Goose revel in barrel-rolling stunts with fighter jets, perhaps what they should be shouting is, "I feel the need—the need to accelerate!" But that doesn't have the same ring to it.

Before Galileo showed otherwise, it seemed sensible that if Earth were indeed flying through space and rotating as it went, we would feel it. After all, if people could feel each bump, wiggle, and jiggle of even the smoothest carriage ride, why would Earth feel any different? We must therefore be stationary in the center of a busy universe, our forebears thought, with the Sun, Moon, and planets revolving around us. What they did not understand was that the larger the vessel, the more stately its movement. And as the ride becomes smoother, the motion becomes less and less perceptible.

As already noted, when a rocket leaves Earth, it receives a free boost equal to the speed of Earth's rotation at the launchpad's latitude. At the Equator, that's an extra 1,000 miles an hour in fuel savings. So, why don't we launch rockets from a mountaintop on the Equator, such as the Cayambe Volcano in Ecuador? There, a rocket has the first 19,000 feet for free, plus the boost from planetary spin. Sounds like a good idea. Turns out, though, that the energy costs of hauling supplies up a mountain more than counteract the benefits. Also, launching along an ocean's eastern shore—for instance, from Florida's Cape Canaveral—offers the convenience of a downwind trash bin for failed launches and discarded first-stage boosters.

Geopolitical factors also influence the locations of launch sites. The European Space Agency launches most rockets from the northern coast of South America, in the French territory Kourou, in French Guiana. At only 5° north latitude, with an expanse of ocean to its east, it's a near-perfect site for a spaceport.

NEWTON, AN APPLE, AND A CANNONBALL

The classic tale of how Isaac Newton discovered gravity is one of the most widely perpetuated—and embellished—stories in the history of science. Maybe you're already familiar with it.

According to the legend, the concept of gravity occurred to

THE CORIOLIS FORCE

Anything that isn't strapped to Earth—say, the air, the ocean, an airborne soccer ball after a kick—experiences and responds to Earth's rotation. This phenomenon is called the Coriolis effect.

Imagine a puffy cloud north of the Equator making its easterly journey when a meteorological low-pressure system emerges directly to its north. The cloud will tend to move toward the low. But during the journey, the greater eastward speed it started with will cause the cloud to overtake the low (which is itself in motion) and end up east of its target. Another puffy cloud—one that starts off north of the low and is moving eastward more slowly—will also move toward the low, but will naturally lag behind and end up west of its destination. To an unsuspecting person on Earth's surface, these curved north-south paths would appear to be the effects of a mysterious force. Yet no true force was ever at work—merely the Coriolis effect.

In the Northern Hemisphere, when many puffy clouds approach a low-pressure system from all directions, you get a merry-go-round of counterclockwise motion, better known as a cyclone. In extreme cases, you get a monstrous hurricane with wind speeds upwards of 100 miles an hour. Down in the Southern Hemisphere, with the same phenomenon at play, cyclones instead spiral clockwise.

Newton in a "eureka" moment in 1666, as he lounged in the shade of a large apple tree in his mother's garden at Woolsthorpe Manor, his childhood home. Trinity College, Cambridge, where he was studying, had sent its students home while the bubonic plague ravaged England. While at Woolsthorpe, Newton noticed

continued on page 56

NOTABLE SPACEPORTS

Svalbard Rocket Range (SvalRak), Norway

SvalRak launches from the northernmost year-round inhabited territory on Earth. No rocket departs Earth farther north than here. As the site yields almost no advantage from Earth's rotation, its launches are suborbital research rockets, used to study weather patterns and such phenomena as Earth's magnetic field and the aurora borealis.

Baikonur Cosmodrome, Kazakhstan

This spaceport was built by the Soviet Union in 1955 as a test range for intercontinental ballistic missiles and is now leased to Russia. The first artificial satellite, Sputnik 1, and the world's first crewed spaceflight, Vostok 1 (Yuri Gagarin's sole spaceflight) were both launched from here. From 2011, when NASA shut down the space shuttle program, through 2020, when SpaceX joined the fray, Baikonur provided the only port of passage for carrying humans to the International Space Station.

Xichang Satellite Launch Center, China

In operation since 1984, Xichang sits at 28° north latitude and has two launchpads. It is a busy spaceport, launching not only meteorological, surveillance, broadcasting, and other satellites but also, in 2007, China's first Moon orbiter.

Ocean Odyssey Launch Platform, Russia

Odyssey was repurposed from a defunct deep-sea oil rig into a bona fide spaceport operating out of the Pacific Ocean. Here, rockets launch almost exactly on the Equator. In 2018 Russia's largest private aviation company bought the floating spaceport. Expect more sea-based launch platforms in the years ahead. The Odyssey itself, however, has been mothballed since landing in Vladivostok in spring 2020.

A NASA suborbital rocket launched in 2022 takes off from Norway's Svalbard Launch Facility.

continued from page 53

an apple fall to the ground. No, it did not bonk him on the head as the legend would have you believe. But he did contemplate the fallen apple, wondering why each one always falls directly downward. Meanwhile, he observed the Moon overhead, in orbit around Earth, and wondered if there was a connection between the two. Where did the influence of gravity stop? Your average thinker might find the attempt to unify these objects perplexing. The falling apple collides with Earth and yet the Moon never does—so how could they be subject to the same power? To observe phenomena so different yet connect them at a profound level required the unique genius of Isaac Newton.

> While at Woolsthorpe, Newton noticed an apple fall to the ground. No, it did not bonk him on the head as the legend would have you believe.

In a subsequent thought experiment, described in his 1687 book *Philosophiae Naturalis Principia Mathematica (Mathematical Principles of Natural Philosophy),* Newton explored and calculated the concept of Earth orbit well before anyone took the idea of leaving Earth seriously. A thrown stone, he knew, would always free-fall toward Earth, but would land farther away with each increase in the strength of the throw. Of course, anyone who has ever hurled a rock realizes this—but Newton took it a step further.

Imagining a cannonball launched horizontally at different speeds, he pondered what would happen as the initial velocity of each cannon launch increased. The stone ball would travel farther and farther, he realized, until at some point it would need to follow the curve of Earth. Not only that: If the cannonball were propelled at a velocity high enough, it would circle Earth completely and hit him in the back of the head. If he ducked out of its way, it would continue its free-fall course, never hitting Earth. At

FALLING THROUGH THE CENTER OF EARTH

K ids sometimes joke about digging a hole through Earth, jumping in, and crawling out on the other side of the world. In the United States, for some reason, they tend to think they'll arrive in China. If they calculated correctly, Americans would actually end up in the southern Indian Ocean.

But what would actually happen? If you could somehow jump into an empty tunnel that extended through the entire Earth, you would continuously gain speed all the way to Earth's 10,000°F core. There you would vaporize, bringing your experiment to a swift end. Ignoring that minor complication, you'd emerge from your trip with a new understanding of the relationship between mass, weight, and gravity. As you fell, and the amount of mass between you and Earth's core lessened, the force of gravity—and consequently your weight—would have lessened too. By the time you arrived at the center, you'd weigh exactly zero.

For these purposes, let's ignore any air resistance you might encounter while plunging down your tunnel. At the center, where you weigh nothing, you're also traveling at maximum speed, so you'll shoot by it and plunge into the other half of the tunnel. Now, gravity tugs against you, slowing you down. Since your journey is gravitationally symmetrical, you'll reach the other end of the 8,000-mile tunnel about 45 minutes after your initial jump, with no remaining speed. Unless you've pre-enlisted a friend (or a fish) to grab you out of the tunnel, gravity will pull and you'll yo-yo back to your original jumping point. That round-trip journey, by the way, will take the same amount of time as one complete low Earth orbit—same as the International Space Station's orbit. Not a coincidence; that's just the way gravitational physics works.

that magic speed, the ball would be falling to Earth at exactly the rate that Earth's round shape curved away from it. We commonly call this state of motion an orbit.

If Newton's cannonball were launched even faster, there's a speed at which it leaves Earth's gravity altogether, achieving escape velocity on what came to be called a hyperbolic trajectory—a concept not yet encountered in Newton's time. In that case, the Moon and the apple undergo identical experiences; the Moon just happened to have a sideways speed that the apple did not.

The International Space Station, which crosses above our heads every 90 minutes, is really just in free fall around Earth at the rate Newton predicted hundreds of years earlier, before the technology existed to test it. Instead of crashing into Earth, the ISS misses the ground over and over and over again. As a result, it remains about 250 miles above Earth's surface, maintaining its orbital speed of slightly more than 17,000 miles an hour. That's almost five miles a second.

Think about what five miles means at ground level. That's 73 times the length of an NFL football field. You could briskly walk it in an hour and a half. A car traveling the speed limit does it in five minutes. A high-speed commercial jetliner can do it in half a minute. The ISS does it in one second. That's how fast objects in low Earth orbit speed across the sky above, and how fast they must move to keep missing Earth on each pass.

And so, astronauts are weightless in orbit—not because space has a magical property that erases the force of gravity, but because they remain in free fall in an orbit Isaac Newton first described almost three and a half centuries ago.

ROCKETEERS IN PEACE AND WAR

The first person to seriously propose a method by which humans could reach the Moon was American engineer Robert Goddard.

HOLLYWOOD SCIENCE

GRAVITY IN SPACE

n the 2019 film *Ad Astra,* every scene inside any space vehicle shows people floating weightlessly. One oversight: The ships are continually firing their engines. The filmmakers could have saved a large chunk of their special-effects budget had they known that when a spaceship fires its engine and accelerates through space, the resident astronauts no longer experience free fall and instead feel the artificial gravity they just created by accelerating.

Back in the days when we sent astronauts to the Moon, their spacecraft had to first accelerate to reach escape velocity. Thereafter, it simply coasted. In other words, the astronauts remained in free fall for most of their trip—not because they were in space, but because they were first in free fall around Earth and then, after the rockets were fired again for what NASA calls a trans-lunar injection, they were in free fall toward the Moon. Moon shots are not currently designed to do this, but if they were equipped with boatloads of fuel and fired their engines continuously, they could accelerate at 32 feet per second, each second, and experience precisely 1 g, or the gravity equivalent at Earth's surface.

A steady 1 g acceleration to the Moon will get you there in two and a half hours, although on arrival you'd whiz by your destination at 55 miles per second. To prevent that, when you're halfway there, you could fire your engines in the opposite direction. You'll then slow down and safely arrive at the Moon after three and a half hours. But at no time will you be weightless during the trip.

In 1926 he launched the first ever liquid-fueled rocket. It was a rickety contraption about 10 feet tall that flew for just two and a half seconds before sputtering back to the ground 184 feet away. Much like the Wright brothers' first successful airplane flight

BALLISTIC BASEBALLS

Have you ever peered over the edge of a high window or balcony and wondered how far away a baseball might land if you hurled one? Would it land on that rooftop over there or that stop sign down at the street corner? If you succumb to this urge, grab a baseball, and throw it with all your strength—you have launched a ballistic missile. And an unassuming passerby would not fare well if they wandered into the ball's ballistic path. The word "ballistic" simply means "under the influence of gravity," and "missile" means anything launched at a target. So, a baseball game, tennis match, football scrimmage, shot-put meet, or any other competition involving targeted flying spheroids is a perfectly legal ballistic missile contest.

SpaceX's Falcon 9 rocket takes off from Launch Complex 40 at Cape Canaveral Air Force Station in Florida.

two decades earlier, Goddard's rocket would beckon a new age of innovation and exploration—and along with it, a new breed of warfare. By the time America entered World War I, in fact, Goddard himself was thoroughly aware of a rocket's military possibilities.

Fast-forward to World War II. On September 8, 1944, a 46-foot-long rocket, loaded with 2,000 pounds of explosives, rose above Earth's atmosphere and fell back to Earth on its target—an ordinary street of a Paris suburb, where it killed six people and maimed many more. It was the world's first long-range, ballistic missile attack, masterminded by the German aerospace engineer Wernher von Braun, who had enthusiastically studied and mimicked Robert Goddard's designs. Having joined the Nazi Party in 1937, Von Braun (in his own words) "fared relatively rather well" as Hitler, eager to prove the technological prowess of Germany, feverishly poured money into his rocket designs.

Fixated on his objective of launching rockets into space, Von Braun in effect accepted—or simply detached himself from—the consequences wrought by his scientific triumph: the first successful launch of a human-made object into space. After the first V-2 dropped onto that Parisian street and two more dropped near London the same day, Von Braun supposedly remarked, "The rocket worked perfectly except for landing on the wrong planet."

Once launched, the V-2 rocket relied entirely on gravity to reach its target. It was a triumph of ballistics. The most terrifying aspect of the V-2 wasn't the quantity of explosives massed within its nose, but rather its unprecedented supersonic speed. The devastating impact from something traveling at such high velocity rendered the onboard bombs a gratuitous fear tactic. Remember, the asteroid that led to the extinction of all the big dinosaurs wasn't toting a single bomb.

Someone standing on the road might have seen a V-2 missile heading at them, but they would never have heard it coming. And even if they could somehow have identified the flying object silently hurtling toward them, it would have been too late. Von Braun's supersonic, suborbital missiles would both extinguish the lives of thousands of innocent people and lay the groundwork for human space travel. Like many tales of marriage between science and warfare, technological advancement is often funded by its promise for destruction.

Although the V-2 rocket itself was a lethal weapon, historians estimate that far more people died building it than died from its strikes. When World War II caused a labor shortage, preventing Hitler from producing his new weapons as speedily as desired, Von Braun turned to prisoners for slave labor. Inmates of concentration camps from across Europe were brought to factories and forced to assemble the weapons in overcrowded underground tunnels where few survived the thirst, hunger, disease, cold, exhaustion, and Nazi brutality. Those deemed unfit to work were sent to death camps.

At least ten thousand people, and perhaps twice that, died constructing the V-2 missile and its predecessors under the most wretched conditions imaginable.

After surrendering to American troops at the end of World War II, Von Braun shifted his career to the United States. At the start of the Korean War, he and his team were transferred to Huntsville, Alabama, where he developed missiles at the U.S. Army's Redstone Arsenal and soon became director of the new NASA Marshall Space Flight Center there. His crowning

continued on page 66

Dr. Robert H. Goddard stands beside a liquid oxygen–gasoline rocket that fired on March 16, 1926, in Auburn, Massachusetts, and flew for 2.5 seconds, climbed 41 feet, and landed 184 feet away.

HUMAN COMPUTERS

All manner of mathematicians and engineers were required to solve the orbital and reentry problems presented by space travel. Key to this effort were the human computers—a team of women enlisted by NASA's predecessor, the National Advisory Committee for Aeronautics, founded in 1915. Their job was to carry out, by hand, meticulous calculations critical to human spaceflight. *Hidden Figures,* the popular 2016 film based on the book of the same name by Margot Lee Shetterly, highlighted three of these trailblazing mathematicians: Katherine Johnson (pictured at right), Mary W. Jackson, and Dorothy Vaughan.

One scene portrayed a famous historical moment between Katherine Johnson and John Glenn, the first American to orbit Earth. In the late 1950s, NASA began using electronic computers to calculate flight trajectories, but as with any new technology, plenty of people didn't want to rely solely on these unfamiliar machines, especially when an equation's accuracy meant life or death. Before Glenn's historic flight into Earth orbit in 1962, he demanded that a human computer verify the electronic calculation of his trajectory. By confirming the electronic results via her mechanical desktop calculator, Johnson helped increase trust in the new technology. And in the category of better late than never, in 2015, five years before her death at age 101, Katherine Johnson received the Presidential Medal of Freedom. In 2019, the street in front of NASA's headquarters in Washington, D.C., received a new name: Hidden Figures Way. And in 2021, headquarters itself was renamed Mary W. Jackson NASA Headquarters.

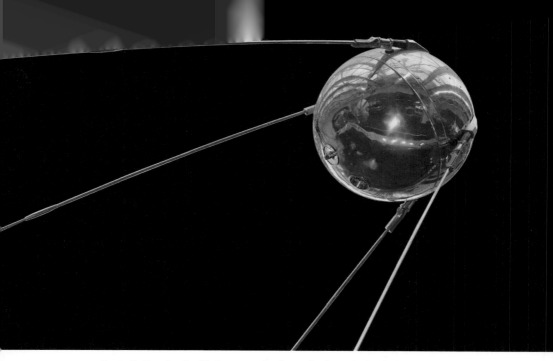

Sputnik, Russian for "fellow traveler," was the first artificial object to orbit Earth. The Soviet Union launched it on October 4, 1957.

continued from page 63

achievement was the Saturn V rocket that carried humans to the Moon.

ATTAINING ORBIT

Although the V-2 rocket was the first human-made object to reach the Kármán line, it never achieved sufficient velocity to reach orbit and thus remained a suborbital triumph.

The first artificial object to orbit Earth was a small satellite launched by the Soviet Union on October 4, 1957—a 24-inch metal ball affectionately named Sputnik, Russian for "fellow traveler." Despite the tender nomenclature, the launch sowed anxiety and terror around the world. Most important, it birthed the space race that further entrenched the United States and the Soviet Union in the Cold War. The successful launch of Sputnik 1 made clear which major power held the lead in technology and weaponry.

To punch Sputnik through the atmosphere and deploy the tiny gleaming globe into orbit, Soviet engineers strapped it atop an R-7 rocket, the world's first intercontinental ballistic missile—20 times more massive than the V-2. It was not lost on anyone that a nuclear weapon could just as easily have sat in place of the beach ball–size satellite. A mere three and a half years after Sputnik 1, in an equally awe-inspiring first, a slightly modified version of the R-7 would launch cosmonaut Yuri Gagarin into orbit: the first human, but only the fourth mammalian species, after dogs, mice, and a guinea pig. Half a year later, a chimpanzee would join the roster.

Aboard Sputnik 1 was a simple radio transmitter that operated on a meager one watt of power—less than the energy consumption of your smartphone in low-power mode. In 1957, that was just powerful enough to emit a beeping radio signal, continually alerting ham radio operators of its peaceable presence. In deep twilight, with clear skies above, anyone under its orbital trajectory with a sharp eye or a pair of binoculars could observe the unfamiliar reflective sphere among the usual tapestry of stars, foreshadowing the thousands of objects now cluttering our once untarnished view of the cosmos.

SPACE JUNK AND THE KESSLER EFFECT

Since that first satellite launch in 1957, humans have pioneered multiple ways to cram as many of them as possible into Earth orbit. With the miniaturization of hardware, satellites have become smaller, cheaper, and ever more indispensable to science and society. The population of satellites in Earth's orbit is growing exponentially—from about 50 deployed in 2000, to about a hundred in 2010, to well over a thousand in 2020, and more than two thousand in 2022. The vast majority of these

continued on page 70

TYPES OF COMMON ORBITS

Low Earth Orbit (LEO)
The region classified as low Earth orbit, up to 1,200 miles above sea level, is populated with the vast majority of satellites. All of our most familiar objects orbiting Earth are found there, including the Hubble Space Telescope and the ISS. At that elevation, full orbits last about 90 minutes, endowing every 24-hour period with 16 sunrises and sunsets. Although LEO is by far the most popular orbit, many satellites inhabit higher regions above Earth, at altitudes better suited for their tasks.

Medium Earth Orbit (MEO)
Extending more than 20,000 miles above LEO is the domain of medium Earth orbit (MEO). America's Global Positioning System (GPS) satellites reside in semi-synchronous orbit at about the midpoint, an altitude at which they complete an orbit every 12 hours, arriving twice each day in the same spot above Earth. GPS satellites send radio signals down to receivers on the ground, including your cell phone, to pinpoint exact locations and distances on Earth's surface. Next time you match with a Tinder date in a radius of five miles or less of where you're sipping coffee, you can thank Isaac Newton (who himself seems to have avoided any romances at all) for his contributions to modern mating habits. Many of the world's other navigation satellites live in the same orbital neighborhood as GPS.

Geosynchronous Orbit (GEO)
Beyond MEO, at 22,200 miles above sea level, satellites circle Earth's Equator every 23 hours, 56 minutes, and 4 seconds—the precise duration of one Earth rotation in space. Although they appear to hover in the same place above Earth's surface, to accomplish this feat they're actually hurtling through space at 7,000 miles an hour, in lockstep with Earth's spin.
 But what about Earth's tilt? A geosynchronous satellite will

sway in the sky as Earth cycles through its seasons. A special case of the geosynchronous orbit—geostationary orbit—occurs in the belt above the Equator where Earth's tilt is near zero. Nearly all the satellites that populate this narrow, highly desirable orbital region facilitate communications, TV broadcasts, and weather forecasting. Occasional spy satellites hang out here too.

Polar Orbit

A specific type of LEO, a polar trajectory runs perpendicular to the Equator. Satellites on this path cross over both the North and South Poles with each sweep. Because Earth rotates within their orbit, a polar satellite, unlike any other, will eventually see the entirety of Earth's surface—ideal for monitoring countries on the other side of the globe.

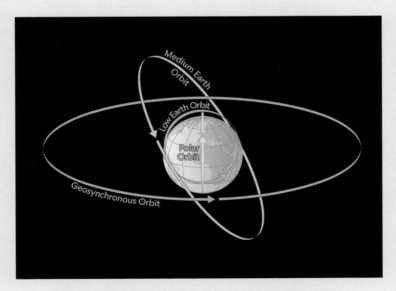

A visual example of satellite orbit trajectories around Earth:
low Earth orbit (red), medium Earth orbit (green),
geosynchronous orbit (blue), and polar orbit (yellow)

continued from page 67

belong to Elon Musk's SpaceX and its Starlink operation. Launches often put multiple satellites in orbit simultaneously. Within mere decades, our once pristine aerial ocean has become a hazardous speedway.

Satellites tend to have a functional life span of 10 to 15 years, but governments and private companies such as SpaceX send them into orbit without any plan to deorbit them once inoperative. The defunct satellites remain ensnared in the orbital current, like vehicles zooming down a highway with drivers asleep at the wheel. In lower orbits, they eventually surrender to atmospheric friction and gravity in a meteoric blaze. In higher orbits, however, where they encounter far fewer atmospheric particles, they can remain to circle Earth for eons.

In addition to all the forsaken, formerly functional satellites, Earth orbit is littered with tens of thousands of pieces of perilous shrapnel. The U.S. Department of Defense's Space Surveillance Network tracks the 30,000 orbiting objects larger than two inches. More than 100 million pieces of additional, untracked

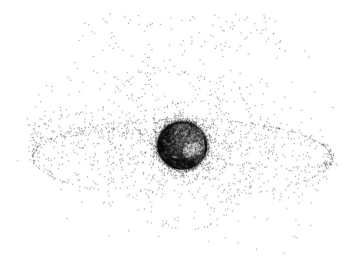

Near-Earth space grows increasingly crowded with abandoned "space junk."

millimeter-size trash are strewn about the sky. At orbital speeds nearly 10 times faster than a bullet fired from an AR-15 rifle, even something as tiny as a paint fleck can wreak havoc on a satellite, a space station, or a spacewalking astronaut.

When two objects collide, the count of broken parts in the wreckage far exceeds the original two. If enough satellites and stray bits remain in orbit unchecked, then catastrophic chaos could ensue. NASA astrophysicist Donald J. Kessler warned of this outcome in 1978. If the fragments of one collision destroy a nearby satellite, and then parts of that now pulverized satellite crash into other nearby satellites, there is a threshold number of satellites above which a self-sustaining cascade of collisions destroys every satellite in orbit. Eventually, countless clouds of shooting debris would preclude safe passage for anything or anyone, and we would find ourselves trapped beneath our self-made junkyard prison.

The 2013 sci-fi suspense film *Gravity,* starring Sandra Bullock and George Clooney, accurately depicted the calamitous consequences of this so-called Kessler effect. What wasn't accurate, however, was Sandra Bullock's bangs, perfectly situated above her brow instead of floating freely in the zero gravity of orbit.

ORBITS AND THEIR DECAY

Three months after the Soviet Union launched Sputnik, the little silvery ball surrendered to our atmosphere, diving back toward Earth in a blaze. Because of its low density, Sputnik disintegrated entirely, like a shooting star, well before it could harm Earthlings below. But its fiery end was not in vain. Scientists gleaned valuable real-world information about how far the outer edges of Earth's atmosphere extend. Even hundreds of miles above Earth, they now knew, air molecules zip about in high enough quantities to cause drag. Over time, as a circling satellite collides with these particles, it loses energy and altitude until it can no longer sustain orbit. If there were no such thing as air resistance, the satellite would orbit forever—unless some other obstruction came along to knock it off course.

Every object circling Earth, especially those in low Earth orbit, including the ISS and the Hubble Space Telescope, must contend with Newton's first law of motion—which tells us that any outside force will change the motion of an object—and must compensate for any stray molecules' frictional effects lest it face the same fate as Sputnik. The ISS dips almost two miles every month. To fight this, it periodically fires rocket thrusters to reboost itself to a higher altitude. When that happens, the astronauts aboard experience a bit of artificial gravity as they're tossed backward in response to the increase in acceleration. To keep up this fight against atmospheric drag, the space station requires constant refueling missions.

THE ROCKET EQUATION

If you want to escape the pull of Earth's gravity altogether and make your way to the Moon, Mars, or beyond, you're gonna need a (much) bigger boat—and a lot more fuel. You'll want something like the Saturn V rocket, completed in 1967 and retired in 1973, which sent the first U.S. astronauts to the Moon.

Designed by Wernher von Braun, the Saturn V stood a towering 35 stories—taller than the Statue of Liberty—and weighed 6.2 million pounds before launch. In the mere 40 years since Robert Goddard sent his 10-foot backyard rocket 41 feet into the air, science and engineering, spurred and funded by war, had advanced at an astonishing pace. Astrophysics became more than looking up. It now included *going* up.

At the top of the rocket was the command module, which held the three astronauts, and the lunar lander. Only the command module returned to Earth from the Moon. Ninety-five percent of the Saturn V's mass consisted of the fuel required to contend with Earth's gravitational embrace.

Newton's third law of motion tells us that for every action there's an equal and opposite reaction. So, if we wish to propel a 6.2-million-pound beast off Earth's surface, we're going to need a powerful action whose force is aimed in the opposite direction. That action was nearly 7.6 million pounds of thrust. The difference between the Saturn V's thrust and weight is the net force upward—more than a million pounds' worth—to launch the thing.

To figure out the amount of fuel required for a mission, rocket scientists must first calculate the weight of the payload and how much fuel will get that payload into space. Here's the problem, though: The weight of the fuel needed to get that payload into space is added to the total payload, and so our rocketeers must now calculate how much more fuel it will take to get that added fuel into space. That newly added fuel weight means we now must add more fuel yet again to make it to space. And so on. Behold the relentless rocket problem, in need of an equation to help us.

The branch of mathematics known as calculus was developed simultaneously and independently by Isaac Newton and Gottfried Leibniz in the 17th century, although components were identified in southern India three centuries earlier. Calculus, exquisitely conceived for this kind of problem, offers the

> Newton's third law of motion tells us that for every action there's an equal and opposite reaction. So, if we wish to propel a 6.2-million-pound beast off Earth's surface, we're going to need a powerful action whose force is aimed in the opposite direction.

continued on page 76

ELEVATORS TO SPACE

I f we could drive a passenger car skyward and *Chitty Chitty Bang Bang* our way to outer space, we could get there in about an hour's time at a normal highway speed (though not in Los Angeles). Space isn't so very far away. However, the oppressive ocean of air and the obstinate force of gravity keep most of us—with the exception of astronauts and billionaires—submerged down here.

Arthur C. Clarke's 1979 novel *The Fountains of Paradise* envisioned a future where access to space meant a cheap and easy trip on a space elevator rather than aboard a rocket. Since then, scientists and novelists alike have explored the science behind what the engineering of such a contraption might require. A space elevator (artist's conception, opposite) combines the concepts of centrifugal force, gravity, and orbital speed. In theory, we could build a massive object (or lasso an asteroid) beyond geostationary orbit to act as a counterweight to a cable, or tether, attached to the ground. If tethered at the correct altitude, the outward, upward centrifugal force of the station or asteroid would perfectly counterbalance the downward pull of gravity at the other end. The cargo compartment—or elevator car—would travel along the tether into the heavens.

Fuel constitutes an overwhelming proportion of a rocket's weight—needed not just to reach a destination but also to carry that fuel and so on. A space elevator, however, renders the fuel issue irrelevant. Maybe we could use solar power for the ascent. The cost per pound to take objects into space might become affordable or, eventually, even negligible.

Here's an additional possibility: At the point of geostationary orbit along the elevator, we could build a space station to serve as a fuel dock, a spaceport to deploy new satellites, or even a spaceship harbor. Any rocket stopping by for coffee could simply zip back into space without ever needing to contend with the tyranny of the rocket equation.

Plenty of obstacles stand in the way of building a structure of this size, of course. Most obviously: The tether, extending tens of thousands of miles, would need to be strong enough to support its own weight. No known substance with such a capacity yet exists, although current research on carbon nanotubes suggests key steps in that direction. Additionally, the tether would need to endure weather from Earth, solar flares, and collisions with space junk or actual satellites at high orbital speeds.

Yet operational space elevators are currently on the drawing board in multiple countries. Whoever builds the first space elevator—or, as China calls theirs, space ladder—will launch a whole new era of mass transit and space exploration.

continued from page 73

means to acquire our needed equation. Although many people independently derived the math, its application to our current problem is more closely associated with the Russian scientist Konstantin Tsiolkovsky, who first devised an equation in the context of space travel to other worlds.

Tsiolkovsky's rocket equation tells us that the fuel needed to lift a payload of a given mass grows exponentially for every extra pound of payload. To gain a modicum of terrestrial insight into this challenge, imagine you wanted to drive your car across the continent—thousands of miles—on one tank of gas. You can't. Your gas tank isn't large enough. So you need a much, much larger gas tank—so large that the weight of the car becomes primarily the weight of gas, greatly reducing your gas mileage, and requiring an even larger gas tank to compensate.

Nowadays an aerospace company might charge $10,000 to put one pound of anything into orbit, although markedly cheaper alternatives do exist, owing to the commercial efforts led by SpaceX to reuse space components. The astronomical expense, especially in the early days of the space program, is why astronauts were trim, and all their electronics were miniaturized. In fact, smartphones and other innovative electronics that now fit in our pockets are the descendants of shrunken technology pioneered during the early space race.

A LESSON FROM METEORITES

So, we now understand a little of what it takes to reach Earth orbit and journey to the Moon. But what about coming home? Astronauts, after all, sign up for a round-trip, not a one-way ticket.

To safely slow an orbiting rocket ship from 17,000 miles an hour to zero requires a braking system. A simple solution might be to fire rockets in the opposite direction (in accordance with

Newton's third law) until it slows to a controlled crawl. But that would require an obscene amount of fuel—exactly the amount needed to launch the ship from Earth in the first place. And because filling stations in orbit or on the Moon don't yet exist, you'd have to carry all that fuel from the start of the journey. The rocket equation is not your friend here.

What's the next best option? Look to shooting stars—meteors—for a clue. When these kinetic space missiles enter Earth's atmosphere, they burn up in what appears from Earth to be a dazzling detonation in the sky. What a shooting star really displays, though, is a progressive exchange of energy. A meteor's kinetic energy converts to heat when the thing encounters the friction of Earth's atmosphere—and sometimes drops a few rocky bits of itself on our planetary surface to become what we call a meteorite. (It's the same conversion that happens when we rub our cold hands together, creating friction that generates heat.)

These chunks of rock travel through the vacuum of space and collide with Earth's atmosphere at speeds upwards of 30,000 miles an hour. They traverse Earth's exosphere and thermosphere without much hindrance, because there isn't much air in these layers, but once the space rocks tear through the mesosphere, we get a light show. In addition to the intense friction produced by the encounter, the gas molecules ahead of the moving meteor quickly compress. And since compressed gas can get as hot as 3000°F, the scorching air vaporizes the meteor before it turns into a meteorite.

Thankfully, most meteorites are no bigger than the size of a pea. Asteroids, their larger cousins, however, can (and do) plunge through the atmosphere and gouge craters in Earth. One famous incident 65 million years ago wrote the final chapter of Earth's then ruling reptilian population.

A space capsule reentering Earth's atmosphere behaves like a giant meteor. Rocket scientists, under the tyranny of the rocket

equation and forced into resourcefulness, saw this not as a problem but as an opportunity. The air offers brake pads for free. The capsule's heat shields work much like any friction brake on Earth. Just as the rubber toe stop of your roller skate converts velocity into heat—through friction with the pavement as it scrapes against it—Apollo-era capsules were clad with a special ablative resin to absorb heat and that burned and peeled away as they plowed through the atmosphere. The more particles the spaceship bashes into, the more heat created in exchange for its kinetic energy—its velocity—and the more the craft slows down. As long as the heat shield continues to wick away the heat, the ship will decelerate without any heat damage.

A space capsule reentering Earth's atmosphere rips through air molecules at thousands of miles an hour. The intense frictional energy sets the protective heat shield ablaze.

Perhaps a better word for heat shields would be "aerobrakes." (After all, we don't call roller skate toe stops "pavement shields.") The now retired NASA space shuttle, which could decelerate from 17,000 miles an hour to zero in half an hour, involved a next-generation heat shield made from a substance called aerogel, which rapidly absorbs and reradiates heat like no other substance. If you take a blowtorch to a sample of the stuff, in the time it takes to put down the torch and pick up the sample, it will have already cooled to room temperature.

These aerobraking heat shields were brilliantly combined with the aerodynamic glide of an airplane, precluding the need for the vessel to be fished out of the ocean. After descending from Mach 25 (25 times the speed of sound) to Mach 1 (the speed of sound), the shuttle's stubby wings could generate lift and drag, just like those of an ordinary airplane, allowing a graceful glide to a gentle halt.

ONWARD TO DEEP SPACE

Five years after the last Moonwalk, humanity set course on its greatest journey into the cosmos, yet not one person was on board. Instead, two humble space probes named Voyager 1 and Voyager 2 carried testimonies to the existence of an intelligent species inhabiting the third rock from the Sun.

Since their launch in 1977, the Voyagers have soared ballistically through interplanetary, and then interstellar, space with their famous golden records in tow. These audio and visual discs, loaded with sounds, songs, greetings, and artwork from around the globe, serve as declarations of our presence to any other intelligent life who might find them. The entire Voyager mission itself, as well as its cargo of technology and culture, encapsulates humanity's fantasy of ascending upward, outward, and onward into the universe.

From Pascal's ascent up Puy de Dôme, to Glaisher and Coxwell's helium balloon brush with the stratosphere (and with death), to Gagarin's orbit around Earth, humans finally cracked open the sky and opened a passageway out into the cosmos on our continued journey to infinity and beyond.

Daedalus would be proud.

This poster is an homage to the Voyager spacecrafts' greatest discoveries, including the volcanoes on Jupiter's moon Io, the nitrogen atmosphere of Saturn's moon Titan, and the cold geysers on Neptune's moon Triton.

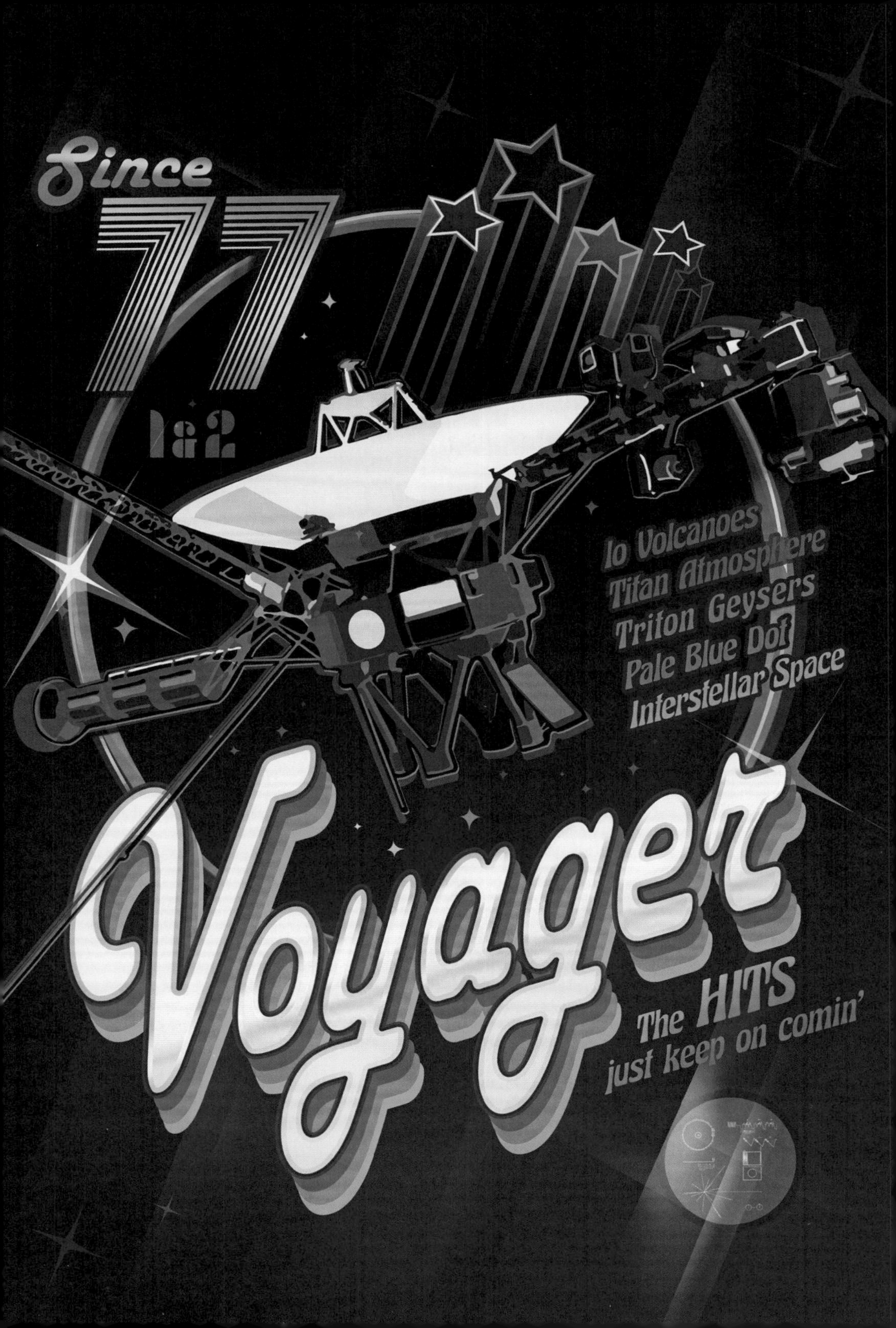
Since
77
1&2
Io Volcanoes
Titan Atmosphere
Triton Geysers
Pale Blue Dot
Interstellar Space
Voyager
The HITS
just keep on comin'

PART 2

TOURING THE SUN'S BACKYARD

"Exploration is in our nature. We began as wanderers, and we are wanderers still. We have lingered long enough on the shores of the cosmic ocean. We are ready at last to set sail for the stars."
—Carl Sagan, *Cosmos*

PRIMVM MOBILE

CRISTALLINE
FIRMAMENT

FIER
AER
YEARTH
WATER

CŒLIFER ATLAS

Hic canet errantē Lunam, Solisq; labores
Arcturūq, pluuiasq; hyad.gēinosq; triōes

I D

The exploration of our solar system is a story with many beginnings, but without end. It is a story of old ideas and assumptions, discarded and replaced by the once unthinkable. From the moment Earth was dethroned from its position as the perfect, unmoving center of the universe, new technologies and advances in mathematics began recasting our cosmic identity into a tale incalculably more thrilling, improbable, and humbling than ever before imagined. In this section, we move beyond our atmosphere to the larger solar system and the distinctive objects whirling around in it, starting with the Sun itself and continuing on to the objects held in its orbital embrace: the innermost rocky planets, monstrous gas giants, mysterious ice giants, and alluring moon worlds among them. To explore the solar system is to recall the histories and fantasies of ancient astronomers—the first scientists who dared suggest a model of the universe—and to marvel at the ongoing missions that uncover ever more of its mysteries.

Earlier astronomers, philosophers, and science fiction authors speculated about the topography and life-forms that might be found on the other planets: beautiful women inhabiting

Atlas bearing the heavens in the form of an armillary sphere,
a model of objects in the sky centered on Earth
(the Ptolemaic system, 1531), compiled by William Cuningham

PREVIOUS PAGES: An optical photograph of radiating coronal streamers
around the Sun, taken in India during a total solar eclipse
on February 16, 1980, using a special camera and filter

the lush jungles of Venus, or intelligent beings who dug complex canal systems into the surface of Mars. Until well into the 20th century, everything we knew about the composition of planets derived from what our eyes could discern at a distance through telescopes and, later, spectroscopy.

Pioneered in the 19th century, spectroscopy—among the most important techniques of modern astrophysics—can determine motions, temperatures, rotation rates, and especially the chemical properties of objects by analyzing their absorbed, radiated, or reflected light. It works for the Sun itself, other stars, gas clouds, and the surfaces and atmospheres of planets, moons, and comets. It even works for entire galaxies. So consequential is the analysis of spectra that its application to astronomy birthed the word "astrophysics" as well as the research publication *The Astrophysical Journal,* inaugurated in 1895 with the subtitle *An International Review of Spectroscopy and Astronomical Physics*. The new science obliterated prevailing assumptions, typified in what now seems a boneheaded comment by 19th-century French philosopher Auguste Comte:

> On the subject of stars, . . .[w]e shall never be able by
> any means to study their chemical composition . . .
> I regard any notion concerning the true mean tem-
> perature of the various stars as forever denied to us.

Before the space race, spectroscopy afforded good but limited information on the composition of Venus's and Mars's atmosphere and surface. Once humankind had the technology to launch probes to other worlds and collect actual samples, we could begin to piece together the history and mysteries of our solar system. Although humans have not set boots on another celestial body since Apollo 17 landed on the Moon in December 1972, the miniaturization of electronics and recent advances in computing and robotics have empowered space probes, planetary rovers, and a Martian helicopter to become humanity's proxy explorers. Let's see what they have discovered—and what unsolved mysteries they have posed.

OUR SUN

Caught in the orbital carousel around our own Sun, eight planets, hundreds of moons, and countless comets and asteroids whirl and twirl on paths choreographed by the forces of gravity. All these objects share a common birthday. Four and a half billion years ago, a star in the Milky Way galaxy exploded, sending out shock waves in its violent death. Triggered by those waves, a nearby cloud of gas and dust, composed of mostly hydrogen, some helium, and a smidgen of other elements collapsed into a flattened nebula—a stellar and planetary nursery. The collapse continued until pressure and gravity caused more than 99 percent of the nebula's mass to coalesce into a dense, amorphous heap. At the core of this central blob, the pressures and temperatures soared so high that hydrogen nuclei fused, emitting tremendous energy and halting any further collapse. Our Sun was born.

Thermonuclear fusion—the contained nuclear bombs that continuously detonate within the Sun's hot, dense core—is the only defense against its own gravity, the only thing preventing its collapse. The crushing pressure and 27 million degrees Fahrenheit temperatures persuade hydrogen nuclei—all positively charged—to overcome their natural repulsion from one another and combine into helium atoms with slightly less mass than the sum of the hydrogen atoms it started with. The "lost" mass is emitted as energy, as prescribed by Einstein's most famous equation, $E = mc^2$. Energy (E) sits on one side of the equal sign, and mass (m), along with the speed of light squared (c^2), sit on the other. As the Sun converts 600 million tons of hydrogen every second, it produces just enough outward pressure to counterbalance its urge to collapse. For about five billion more years, while our star continues its stable life, Earthlings will enjoy a steady stream of its radiative energy.

A WINDY STAR

Strange as it may sound, all stars have an atmosphere of their own. The outermost layer of our Sun's atmosphere is called the corona—a word most of humanity became familiar with during the 2020s as a coronavirus swept through the world. Both the Sun's outer layer and the virus particles that caused the latest global pandemic were named after the Latin word for "crown" because of their resemblance to the spiky headpiece.

The Sun's corona can reach a couple million degrees Fahrenheit—a temperature so hot that electrons and protons are stripped from its host atoms and stream off into space at nearly a million miles an hour. This continuous torrent of charged particles, called the solar wind, extends billions of miles in every direction, enveloping every object in the solar system. Its limits of influence mark the official edge between our solar system and interstellar space.

While the solar core keeps busy nuking itself, the other 99 percent of the Sun responds to the resultant heat and pressure. Forces deep within the core churn the Sun's surface and mobile magnetic field. Every 11 years or so, the Sun's magnetic poles flip completely. The flip itself is not sudden or calamitous; it simply marks the transition from one solar cycle to the next. Initially there's a period of calm, called the solar minimum, which gives way to a period of increasingly intense sunspot activity, peaking at what we call the solar maximum. That's when calamity can ensue.

It plays out this way: Roiling blobs of electrically charged plasma are occasionally belched out as coronal mass ejections (CMEs). The sudden and violent eruption of a CME can send

A solar flare as seen in a bright flash, captured by
NASA's Solar Dynamics Observatory in 2022

shock waves through the more slowly moving solar wind; if directed at Earth, a CME can be felt and seen as a solar storm, with devastating potential. Fortunately for us Earthlings, though, the magnetic field engulfing our planet offers protection from the onslaught of ionizing radiation. Although most of the charged particles are deflected away from Earth, some are funneled into our upper atmosphere. There, they collide with gas particles to bestow some of the greatest light shows on our planet. The northern and southern lights, aurora borealis and aurora australis, appear as gorgeous curtains of green, red, and purple strewn across the otherwise dark skies.

But not all solar weather is so beautiful or benevolent.

The 19th-century English astronomer Richard Carrington spent years mapping the many dark spots seen on the Sun's surface. His technique was clever: He aimed his telescope at the Sun but projected its image back through the eyepiece onto a

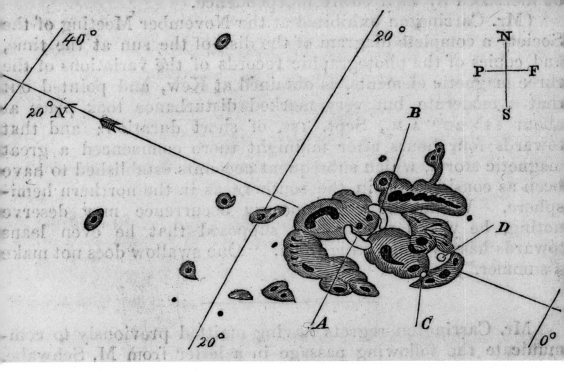

An illustrated plate of a group of sunspots based on observations by the British astronomer Richard Christopher Carrington, from *Memoirs of the Royal Astronomical Society* (1861)

screen against the wall. By studying the image on the screen, he could discern the surface features without harming his vision. (It's ill-advised to look directly at the Sun, especially with a telescope, if you're planning to look at anything else ever again.)

One day in 1859, during the course of Carrington's usually tedious monitoring, bright white lights suddenly danced across the darker spots, only to disappear minutes later. What he had observed was a solar flare—the first ever detected. These flares tend to coincide with CMEs but are a separate phenomenon. Rather than carrying plasma particles, as CMEs do, flares carry only energy, mostly as x-rays, that travel at the speed of light. A solar flare takes 500 seconds to reach Earth because the Sun lies 500 light-seconds away; a CME, on the other hand, could take a day or more to traverse the same distance.

In the case of Carrington's observation, an unseen CME indeed erupted alongside the observed solar flare and had begun its journey toward Earth, where it would become the most powerful solar storm ever recorded. Auroras could be seen across the world as electrically charged particles engulfed the entire globe. Those particles also traveled down telegraph lines. There were reports of electrical shocks striking telegraph operators unconscious, and showers of sparks setting their desks aflame.

If such an event occurred today, given our dependence on a ubiquitous power grid and GPS satellites, apocalyptic chaos could well ensue. Credit and debit cards would instantly stop working. And you can forget about your cryptocurrency. Planes would be grounded, food and fuel supplies would quickly wane, and society altogether would screech to a halt, with nary a radio program or television station to broadcast the situation. Above Earth, any spacewalking astronaut would have no more than a few minutes' warning to seek shelter before the impending fatal radiation would strike.

> If such an event occurred today, given our dependence on a ubiquitous power grid and GPS satellites, apocalyptic chaos could well ensue . . . And you can forget about your cryptocurrency.

Can we defend against such a catastrophic event? Possibly. With better understanding, we may be better able to predict solar cycles and prepare for the weather they bring. NASA's Parker Solar Probe set out to achieve just that, and in 2021, it flew through the Sun's corona and became the first spacecraft to brush our star. As with all edges and boundaries, the point of separation between the solar wind and the Sun's atmosphere is difficult to isolate. The Parker Solar Probe helped to further define the fuzzy edge of that theoretical boundary— what astrophysicists call the Alfvén critical surface, where the solar wind particles begin to escape the influence of the Sun's

WHAT COLOR IS THE SUN?

I n grade school, when you drew pictures of the Sun, you probably reached for a yellow crayon. Chances are, every painting, celestial map, old astronomical drawing, or model you've ever seen of the Sun showed a yellow orb. So you may be surprised to learn that the Sun is, in fact, white. It emits every color of the rainbow, and as Isaac Newton demonstrated in the late 17th century, culminating in his seminal 1704 book *Opticks,* an equal combination of all those colors creates white light. The reason the Sun looks yellow is the same reason the sky looks blue.

When the white sunlight passes through the atmosphere, air molecules preferentially scatter the shorter blue waves in all directions, removing a little bit of the energy that would otherwise make it to Earth. Our blue sky is really stolen sunlight. During sunrise or sunset, when the Sun lies low on the horizon, its light travels through nearly 40 times as much atmosphere before it reaches you, compared with when the Sun is directly overhead. The sky turns a deeper and deeper blue as the blue light waves are scattered and rescattered, while the Sun turns more and more amber, adorning the horizon with a gorgeous tapestry of crimson and orange.

gravity and magnetic field as they rush forth into the solar system at large.

Nearly two centuries have elapsed since Richard Carrington first mapped solar flares, and more than two millennia since the Chinese astronomer Gan De first documented spots on the Sun. The behavior of these solar tantrums, alas, remains unpredictable. Why and how do they form where they do? Which ones might unleash catastrophe on Earth's inhabitants? We await the scientific discoveries of missions such as the Parker Solar Probe, which could unlock the necessary knowledge to predict and conquer future storms unleashed by the great nuclear bomb in the sky.

The less than one percent of matter that resisted becoming the Sun 4.6 billion years ago formed every other celestial body in our solar system. While each of our planets and moons sports unique features and surfaces, all of them carry the same array of elements contained in that primordial nebula.

The origin story of the celestial bodies that orbit the Sun remains hotly contested. One point of agreement: The early solar system was chaos incarnate. The particles that didn't merge into the Sun coalesced into chunks called planetesimals; some combined further into huge bodies called protoplanets. This bedlam of jumbled objects collided with one another—knocking some apart and pushing others out of orbit. Some plunged into the Sun, adding to its mass; other objects were ejected from the Sun's gravity entirely. Some of the smaller ejectees became rogue asteroids, destined to become interstellar interlopers, while the larger ones became rogue planets—celestial vagabonds with no star to call home.

Up to 30 planets may have participated in this ruthless game of cosmic billiards. Only eight survived. The survivors' roster includes the four rocky inner planets—Mercury, Venus, Earth, and Mars; the two gas giants—Jupiter and Saturn; and the two ice giants—Uranus and Neptune.

MERCURY, FIRST OF THE ROCKY INNER PLANETS

The four planets closest to the Sun, also called the "terrestrial" planets, each have a large metallic core, rocky surfaces with distinctive topology, and few or no moons. Among them—and really, among all the eight planets that populate our solar

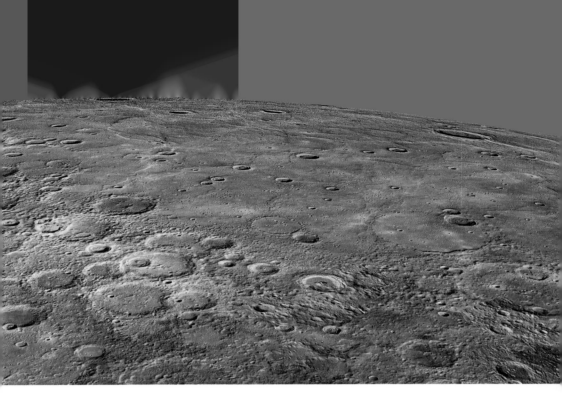

system—Mercury carries the worst reputation. It's the puniest of all and has the most heavily pocked and scarred surface. With its extreme temperatures and embarrassingly thin atmosphere, Mercury is utterly barren and inhospitable. And that's not its worst offense. About three times a year, for a few weeks at a time, this little planet appears to move backward across the sky—a turn of events commonly called Mercury in retrograde. Bewildered by this bizarre phenomenon, ancient humans (and many modern ones as well) blame Mercury's retrograde for their misfortunes.

There's a backstory.

For most of human history, the world was simply how it appeared and felt. The scientific method—the urge to test a hypothesis via repeated experimentation—didn't take hold until the 17th century. That's how Aristotle got away with declaring that heavy objects fall faster than lighter objects, and why people believed him for two millennia. A few simple experi-

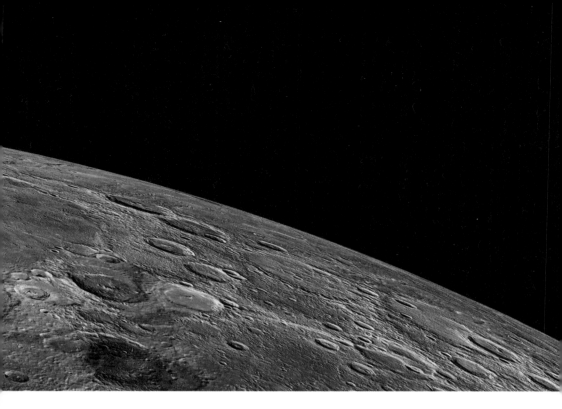

The rainbow hues of Mercury's colorized topography, with purple representing the lowest height variation and white the highest

ments with various-size rocks would have instantly revealed the errors of his thinking.

For a thousand years, Earth was regarded as the unmoving center of the universe, a view Aristotle popularized and Ptolemy later embellished. Everyone regarded this as an obvious feature, because all celestial bodies and constellations appeared to migrate around Earth. This was a given—until the 16th century, when the Polish astronomer Nicolaus Copernicus proposed the radical idea that the Sun, not Earth, lay at the center of the universe and that, along with every other planet, ours revolved around the radiant orb.

To the naked eye, seven distinct lights appear to move like wandering vagabonds across the otherwise harmonious constellations. The ancient Greeks called these bodies *planetes* (wanderers): the Sun, the Moon, Mercury, Venus, Mars, Saturn, and

Jupiter. In the course of one year, the wandering Sun—from the perspective of an earthly observer—appears to slowly pass through 12 constellations. The wandering Moon can be "new" or "full" or something in between during its monthly journey across the sky. Mercury and the other wandering lights appear to intermittently reverse the direction of their paths before reversing yet again across the sky. That's how it looked to the geocentric world, so presumably that's what was happening. Hence the idea of retrograde, as well as other terms still in use today, such as "sunrise" or "sunset"—geocentric terms that pre-date our heliocentric knowledge of our own solar system and the cosmos as a whole.

When Earth was demoted from the center of the known universe to an ordinary planet in orbit around the Sun like everybody else, we recognized retrograde for what it was: a simple optical illusion. Because Earth and Mercury move relative to each other, and both orbit around the Sun, we witness Mercury both in front of the Sun as well as behind it—a perspective our ancient ancestors could not yet reconcile with their worldview.

Just as satellites in low Earth orbit move more quickly than those in higher, farther orbits, planets that orbit closer to the Sun move more quickly than do their faraway cousins. The ancients named the swiftest planetary wanderer after the winged-foot Roman god of speed and travelers: Mercury. One Earth year is a little more than four years on Mercury—which is to say, Mercury revolves around the Sun four times to our one—and, not incidentally, three or four times a year, Mercury appears to go into retrograde. That is, during its orbit, when Mercury passes by Earth and swings to the far side of the Sun, it appears to reverse direction and travel the other way.

You can easily visualize this: Imagine a spinning musical carousel—the kind with painted wooden horses you might find at a county fair. Your friend rides atop one of those horses, circling round and round, as you watch from the crowd. As she passes closest by you and waves, she seems to move from your left to your right. But when she's circled around to the far side of the ride, from your point of view, she is now moving from right to left. You know that the horse did not suddenly break away from the ride and start galloping backward. It's circling the center of the ride, and you are not.

From a Sun-centered perspective, Mercury doesn't move backward either. What we call "Mercury in retrograde" is a function of our perspective on the Sun and the planet, not a reversal in Mercury's orbital direction. There's nothing unusual about Mercury's movement during those times, other than the human urge to assign it meaning—an urge traceable to a time of ignorance, when we all thought the universe revolved around us.

Because Mercury and Venus are closer to the Sun than Earth is, we occasionally lose sight of their path in the glare until they reemerge. That entire time, like the horse on the other side of the carousel, they appear to travel in the opposite direction as before. The planets more distant from the Sun than Earth also seem to demonstrate retrograde motion relative to Earth, but this is dominated by our motion around the Sun, which is faster than theirs.

> If every planet moves retrograde with respect to every other planet in the solar system, then why castigate Mercury?

If every planet moves retrograde with respect to every other planet in the solar system, then why castigate Mercury? The number of times a planet goes into retrograde relative to Earth is directly correlated to its orbital speed relative to Earth. So, Mercury backslides more often each year than all the other planets do. Faraway, slow-moving Neptune goes retrograde only once a year, exhibiting

continued on page 100

HOW EINSTEIN KILLED PLANET VULCAN

You may be familiar with the fictional *Star Trek* planet Vulcan, Lieutenant Spock's home planet. Vulcan's residents are a logical, emotionless alien race with pointed ears and upswung eyebrows. But more than a century before the first *Star Trek* episode aired, Vulcan was thought to be a real planet in our own solar system.

Copernicus's heliocentric model of the universe, Kepler's laws of planetary motion, and Newton's laws of physics had turned the solar system into a knowable place—for the most part. In the mid-19th century, French mathematician Urbain Le Verrier found himself utterly stumped when attempting to account for a pernicious annual shift in Mercury's orbit. Again and again, astronomers tracking Mercury's orbit tried in vain to predict it accurately, even though the orbits of the other nearby planets behaved as expected. Newtonian mechanics flat-out failed. The only way to account for the deviation, Le Verrier postulated, was a hidden, flaming hot planet that lay between Mercury and the Sun, tugging at Mercury and making it do things that made Newton's laws look bad. It was named Vulcan, after the Roman god of fire. This may seem an outlandish claim—after all, if another planet really did exist in the inner solar system, wouldn't somebody have seen it already?

Yet to observe such a planet with the unaided eye, or even with a telescope, would prove a difficult task indeed. The Sun's blinding glare would be prohibitive. The only opportunities to see a planet orbiting that close to the Sun would occur in dusk or dawn twilight, or during a total solar eclipse.

Le Verrier had already garnered great respect for having predicted the existence of Neptune based solely on Uranus's deviant orbit. As a result of his popular acclaim, scientists generally believed him, and the mysterious planet Vulcan was written into the textbooks despite the lack of any reliable observational evidence.

In 1915 Albert Einstein proposed his general theory of relativity, which, among other things, explained the behavior of gravity under extreme conditions—such as very near the Sun. With this new understanding of the universe in hand, Mercury's odd orbital behavior could be explained, forever extinguishing fantasies of a real planet Vulcan.

Leonard Nimoy—as Lieutenant Spock—gives the Vulcan hand greeting.

continued from page 97

backward motion during the half year that Earth spends moving around the far side of the Sun.

So most of the time, at least one planet is going retrograde relative to Earth, yielding no end of excuses for everybody's bad luck. Maybe Shakespeare said it best: "The fault, dear Brutus, is not in our stars, but in ourselves."

EXPLORING MERCURY: GRAVITY ASSISTS

Mercury remains the least explored of the seven classical wanderers. It's not that the tiny, cratered world is uninteresting—far from it—but rather that getting there poses great challenges. In fact, our first Mercury orbiter didn't arrive until 2011—decades after we had landed on the Moon. It's far easier to get to Jupiter, which is about seven times farther from Earth, than it is to get to Mercury. And here's a stupefying fact: It's cheaper and more fuel-efficient to escape the solar system entirely than to land on the innermost planet.

As we know by now, objects in orbit move more quickly the closer they are to the source of gravity. Mercury's speed, combined with its tiny size (not much larger than our Moon), makes it a difficult target to hit. But there's more. A spacecraft that has escaped Earth orbit—in other words, achieved escape velocity—is traveling at least 25,000 miles an hour. As it approaches the Sun, it speeds up from the powerful gravitational pull. Mercury's low gravity means an object must be traveling much, much more slowly than escape velocity to be pulled into its orbit. Without brakes, an orbiter heading for Mercury would be traveling so fast that it would overshoot the planet completely. Obviously, it has to slow down somehow. But the tyranny of the rocket equation strikes again: The spacecraft would need to carry more fuel—more weight—just to brake than it could launch with. And, given

Mercury's nearly nonexistent atmosphere, aerobraking is out of the question.

Faced with this intractable problem, scientists devised an extraordinarily elaborate maneuver. The solution to slowing down without fuel involved a gravity assist—the same trick that sped up Voyagers 1 and 2 as they flew by the planets and escaped the solar system decades ago.

So, how did Voyager do it? The year 1977 brought an uncommon planetary configuration of Jupiter, Saturn, Uranus, and Neptune that only occurs every couple hundred years. NASA scientists knew they had a unique opportunity to borrow energy from the aligned planets to slingshot a spacecraft out of the solar system at more than 30,000 miles an hour, and thus overcome constraints of the rocket equation.

You might think that in a cosmic slingshot, a planet's gravity would pull in an incoming object and spit it out the other side faster than before. But, no. The same gravity that pulls you in also resists your efforts to leave. Remember our thought experiment about jumping through Earth? You can gain speed only until the planet tugs you back toward it, in a perfect and never ending cycle. The net speed gained from gravity alone? Exactly zero.

So where does the slingshot speed come from? If you approach a planet in its orbit from behind, the moving planet will draw you in with it, allowing you to steal some of its orbital speed. Without the gravity assist, Voyagers 1 and 2 would have been stuck in orbit around the Sun somewhere between Jupiter and Saturn.

Using the technique the two Voyagers perfected, scientists figured out how to sufficiently slow down a spacecraft headed toward Mercury. How can you use a gravity assist to brake rather than accelerate? You approach a planet in its orbit head-on rather than from behind, transferring some of your orbital energy onto it. The planet speeds up a teensy bit, and you slow down.

NASA launched Mercury-bound MESSENGER in 2004. Seven years later, after multiple gravity-assisted decelerations around Earth, Venus, and Mercury itself, MESSENGER achieved orbit around its destination. For four years it circled the pockmarked planet, taking thousands of pictures, answering old questions, and posing new ones.

Despite its proximity to the Sun, Mercury is not as hellishly hot as you might think. In fact, it's cooler than Venus. Daytime surface temperatures do reach a scorching 800°F in the sunny spots, but temperatures inside its deep, eternally shadowed valleys plummet to nearly minus 300°F. An almost preposterous consequence of these low temperatures is that MESSENGER detected evidence of water ice locked within the planet's frigid poles.

Although the MESSENGER mission was designed to last only a single year, it surpassed expectations and endured for another three years. But its fuel supply inevitably ran dry, and with no defenses left against gravity's beckoning, it began a collision course with the cratered planet and, in April 2015, left a crater of its own.

The European Space Agency (ESA) and the Japan Aerospace Exploration Agency (JAXA) are now heading to Mercury together to pick up where MESSENGER left off—literally. One objective of the mission, called BepiColombo, is to identify and study MESSENGER's final resting place. The impact likely kicked up terrain and loosened debris from below the surface, stuff that

> Despite its proximity to the Sun, Mercury is not as hellishly hot as you might think. In fact, it's cooler than Venus. Daytime surface temperatures do reach a scorching 800°F, but temperatures inside its deep, eternally shadowed valleys plummet to nearly minus 300°F.

THE FIRST SCIENCE FICTION STORY

n 1608, two years before Galileo would see evidence that Earth moves around the Sun, the great German mathematician Johannes Kepler wrote what some consider to be the first ever piece of true science fiction. He called it *Somnium (The Dream)*.

The story tells of a young Icelandic boy and his mother—an herbalist and witch—who could summon nonhuman beings called daemons capable of traveling back and forth between Earth and the Moon, which the beings call the "island of Levania." One night, the witch asks one of these traveling daemons to describe what the solar system looks like to the residents of Levania. "To its inhabitants," wrote Kepler, "Levania appears to stand unmoving, among the moving stars, no less than our Earth appears to we humans."

To us, this sentence may seem harmless. Today we know that no matter where we stand—on Earth, the Moon, or any other rotating celestial body—all other objects will appear to move around us, not us around them. But to those living in a society committed to the belief that God made Earth the perfect and unmoving center of the universe, Kepler's description was nothing short of radical, even when incorporated into a work of fiction.

Knowing his provocative words could be interpreted as blasphemy, Kepler was careful to distribute his manuscript to only a select few scientific friends. Despite his caution, however, it fell into the wrong hands. A few people perceived condemnable similarities between the main character and Kepler himself. His own mother, accused of being the story's daemon-summoning crone, was tried for witchcraft and imprisoned. Six years of legal battles later, she was exonerated, but she died the very next year. The manuscript remained unpublished during Kepler's lifetime.

Despite the hardships wrought by his novel's implications, Kepler never relinquished his heliocentric worldview.

may offer insights into this secretive planet. Perhaps MESSEN-GER's legacy is not fully written, as new cosmic discoveries may await us in the remnants of its demise.

VIBRANT VENUS

Both Mercury and Venus appear near the rising Sun just before dawn, disappear for weeks or months at a time, and reappear near the setting Sun at dusk. Given their close proximity to the Sun, they never wander too far from it when seen from Earth's sky.

A composite image of Venus's transit (in black)
in front of the Sun in 2004, the planet's first solar transit since 1882

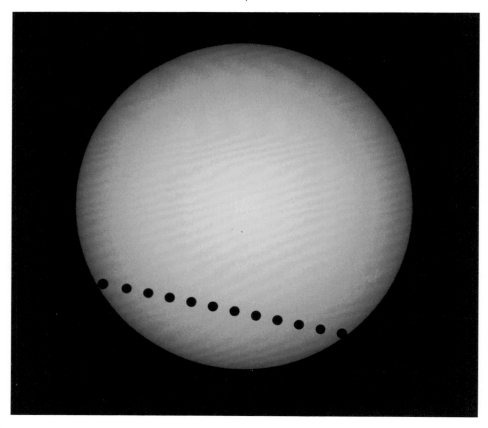

Some ancient civilizations, Greek and Roman among them, held that Venus was in fact two separate stars, each with its own identity. Glowing, silvery Venus, the brightest object in our sky after the Sun and the Moon, appeared to them as both a morning star and an evening star. The ancient Greeks called its evening appearance Hesperus and its morning appearance Phosphoros (in Latin, Lucifer, meaning "Light Bringer"). Eventually, after accepting the two separate manifestations as one and the same object, the Romans renamed the vivid orb after their goddess of love and beauty, Venus.

Although some ancient astronomers regarded the planet's morning presence and evening presence as two different heavenly bodies, many other ancient civilizations—the Babylonians, Aboriginal and Torres Strait Islander peoples, and many Indigenous Central American cultures, especially the Maya—understood that the occasionally disappearing lights emanate from the same object.

PHASES OF VENUS

When Galileo Galilei viewed Venus through his improved version of the telescope in 1610, he witnessed a feature of the vibrant planet that would help overthrow the geocentric universe. Venus, he saw, displays phases just like the Moon. For example, when the Sun was to Venus's left (as seen through Galileo's instrument), the left side of Venus was illuminated while the right side was in darkness. Behold: half Venus. Successive configurations of the Sun and Venus result in Venus displaying the same familiar phases we witness for the Moon. Galileo sensibly concluded that Venus must revolve around the Sun rather than around Earth. It was a nail in the Ptolemaic geocentric coffin and solid evidence in favor of the heliocentric system, nearly 70 years after Nicolaus Copernicus proposed that model.

Though Galileo was eager to claim his new discovery, he was hesitant to announce it immediately. He sent a riddle to his friend Johannes Kepler, encrypted as an anagram. It might seem an odd way to announce a scientific breakthrough, but back in the days before copyright, patents, and peer review, this was a common practice. He wrote, *"Haec immatura a me iam frustra leguntur o.y.,"* which roughly translates as, "These are at present too young to be read by me." Once Galileo was confident in his discovery, he unjumbled the letters to convey his hidden message: *"Cynthiae figuras aemulatur mater amorum,"* meaning "The mother of love emulates the shapes of Cynthia." Venus, of course, is the mother of love in this riddle, and in Roman mythology the goddess of the Moon was often called Cynthia. So, the fully solved puzzle revealed Galileo's discovery: Venus emulates the Moon.

TRANSIT OF VENUS

In 1639, a 20-year-old English tutor prepared to witness a cosmic event that nobody else had ever seen. The young Jeremiah Horrocks, a prodigious reader of the newest texts in astronomy, analyzed the latest planetary tables and calculated that Venus was about to pass directly in front of Earth's view of the Sun. For several hours, a planet we had only before witnessed as a gleaming orb in the night appeared as a tiny black dot moving across our star.

Horrocks safely observed the transit by training his telescope on the Sun and projecting the image onto a piece of paper. He and his companion William Crabtree were the only people on Earth to watch the spectacle. On the basis of his observations, astronomers estimated the distance from Earth to the Sun, the size of Venus, and the size of the entire solar system out to Saturn, the farthest planet then known.

Horrocks died 14 months later. Very few records of his original works and observations survived the centuries since, many lost to the Great Fire of London. We may never know what those records contained or how he may have further shaped the course of astronomical discovery. His partially rescued and posthumously published work proved a mind and life as exceedingly rare and as tragically fleeting as a transit of Venus.

By the time of the 1769 Venus transit, new measuring tools had become available, enabling sky-watchers to accurately derive the Earth-Sun distance and thereby figure out a truer size for the solar system. Astronomers were dispatched to scores of locations across the globe where the transit would be visible, from Norway and Siberia to Baja California and India. Famed and notorious British explorer James Cook, who set sail with astronomer Charles Green aboard the HMS *Endeavour* to document the transit from Tahiti in 1769, captained the best known expedition. This time, the Brits made a visual record of what they saw.

Then, armed with Venus transit data obtained by these simultaneous expeditions, including precise timings and coordinates on Earth's surface, astronomers used plain old trigonometry to determine an accurate value of Earth's distance from the Sun.

While new technologies have superseded the Venus transit as the preferred method for sizing up our place in the universe, these rare events continue to offer valuable opportunities to understand the cosmos. The next one is scheduled for December 2117. By then, if the ambitions of space enthusiasts prevail, humans might be a multiplanetary species, able to witness transits from Earth, the Moon, and Mars.

EXPLORING VENUS

Before the space race, the thick, reflective atmosphere encapsulating Venus shrouded Earth's nearest neighbor in mystery. The

VENUSIANS

Ve-nu-sian \vi-ˈnü-zhən\ adj (1874): of or relating to the planet Venus

We would call an alien from Venus Venusian, but that's only because the medical profession already claimed the correct Latin adjective. If we were to follow the rules of Latin grammar, a person or thing relating to Venus—the planet named for the goddess of love and beauty and all that goes with it—should be called Venereal.

most that Earth-based observations could glean was its size, distance, and the chemical composition of its upper atmosphere. In late 1962, a year and a half after the first human, Yuri Gagarin, was sent into Earth orbit, the first probe to another planet set sail. NASA's Mariner 2 soared past Venus, taking its temperature along the way: a scorching 450°F overall. Later measurements doubled that figure.

Soon the Soviet Union set to work on its Venera program—a series of Venus probes and landers—that would reshape our understanding of the planet next door. In 1967, Venera 4 became the first probe to enter the atmosphere of another planet. In 1970, Venera 7 made the first soft landing on another planet's surface, relaying a surface temperature of almost 900°F and a crushing atmospheric pressure 90 times that on Earth. It also revealed a toxic atmosphere of 97 percent carbon dioxide. Venera 7's successful mission dashed all remaining hopes that Venus could be home to a colony of beautiful women or indeed any other life—beautiful, female, or otherwise. Five years later, Venera 9 took the first ever pictures from the surface of another planet, revealing Venus's rock-strewn desolation—nothing like the tropical ocean planet teeming with life that sci-fi writers had imagined.

A LESSON FOR EARTH FROM VENUS

Before the Mariner and Venera probes returned their data confirming scorching surface temperatures, an American astrophysics doctoral candidate named Carl Sagan predicted that Venus's thick atmosphere could render the planet uncommonly hot and cause a runaway greenhouse effect. Although Earth requires a bit of a greenhouse effect to maintain an environment suitable for life as we know it, there is a point of no return: If an excess of greenhouse gas were to build up, Earth's water would boil off the surface and the now desiccated planet would trap even more incoming and infrared energy under its dense atmosphere.

The definition of a greenhouse gas is that it absorbs and emits infrared while allowing other frequencies of light to pass through. You may be thinking that either carbon dioxide or methane is the most potent greenhouse gas. Nope, neither. Of the major ones, water vapor easily takes first place. And so, when a planet containing water becomes so hot that all its surface water evaporates, becoming atmospheric water vapor, then the greenhouse effect quickly cranks all the way up.

That's the story of Venus. Once a water world, perhaps even lush and teeming with life, as Earth is today, Venus is now a desolate volcanic hellscape with an average surface temperature hot enough to melt lead. Sagan wondered what could happen if our own planet's moderate greenhouse effect increased due to rampant burning of fossil fuels, releasing long-buried carbon into the atmosphere as carbon dioxide. His calculations led him to testify before Congress in 1985 about the dangers of anthropogenic climate destruction.

But could Earth ever really become Venus?

Each year, Earthlings spew more than 50 billion tons of heat-trapping greenhouse gases into the atmosphere. Carbon dioxide, from burning fossil fuels—coal, natural gas, and oil—constitutes three-quarters of these emissions. As the name implies, fossil fuels derive from ancient remains of plants and animals. Despite the logo of a certain large oil corporation, none of the corpses that make up its product were roaring reptiles. Nearly all the oil and most of the natural gas we use to power our cars, planes, and homes comes from ancient microscopic plant and animal organisms called plankton—the phytoplankton that feed on sunshine and carbon dioxide, and the zooplankton that feed on the phytoplankton. Whereas our coal derives from the buried remains of multicellular forest plants.

A recipe for fossil fuels: Collect copious plant or animal matter. Exert relentless pressure for millions of years. Heat until you reach your desired consistency.

Three hundred million years ago, life—bizarre and fantastical life—flourished in the swampy marshes and warm shallow ocean that submerged most of the globe. Fossils from way back then tell a story of a highly oxygenated, humid atmosphere, where hawk-size dragonflies and foot-long scorpions darted through lushly forested wetlands. Billions of trees, some species more than a hundred feet tall, siphoned carbon dioxide from the air and developed strong fibers, called lignin, that supported their weight yet retained flexibility in the wind. But with their thin, towering trunks and surprisingly

Artwork depicting a Carboniferous period landscape of plants and insects some 280 million to 340 million years ago

shallow roots, these giant, fast-growing trees fell over easily—and accumulated quickly in the low-oxygen waters below. Over time, silt and rising seas swallowed the megatons of decaying plant material, compressed by the immense heat and pressure. Their fate: to become carbon-rich coal, fueling human activities 300 million years later. That era of widespread tree deposits is known as the Carboniferous period. Nearly all the coal we use today was created then.

Most of the world's oil and gas derives from around 100 million years after this era, during the reign of dinosaurs in the Mesozoic age, or the age of reptiles. The warmer oceans of this time allowed plankton to flourish. As with the trees of the Carboniferous period, their sunken remains were exposed to extreme pressure and heat over tens of millions of years. Although some natural gas derives from the late stages of coal breakdown, nearly all oil was formed from plankton. These organisms—the single-celled photosynthetic phytoplankton that feed on sunshine and carbon dioxide, and the zooplankton that feed on the phytoplankton—are, like us and all other life, carbon-based. When the complex carbon molecules of life burn, the chemical reaction releases CO_2. A by-product—more useful to industrious humans—is plenty of energy in the form of heat.

Because our fossil-fuel recipe requires millions of years to bake, these fuels are functionally a nonrenewable resource. In other words, the amount we will ever find within Earth during the life span of our species is fixed. During the so-called energy crisis of the 1970s, analysts at the U.S. Energy Information Administration made harrowing predictions about how little oil and gas remained available for human consumption, raising alarm about a quickly dwindling supply.

Meanwhile, humans poured plenty of funding into innovative research on new extraction methods to sustain our growing need (and greed). The post–World War II embrace of hydraulic fracturing (also known as fracking, which was dreamed up in the 1860s) brought formerly difficult-to-reach fuel reserves within reach, and as our supply skyrocketed, fuel costs have generally plummeted. As long as there's funding for new ways to extract fossil fuels, and as long as human societies remain dependent on them to power transportation, manufacturing, heating, cooling, and nearly every other aspect of everyday life, fossil fuels aren't going away anytime soon.

But let's say we do run out of fossil fuels. If humans burned every ounce of coal, oil, and gas sequestered in Earth's crust, the planet would revert to something vaguely resembling its former self of 300 million years ago. Life would still flourish—just not the way it looks today.

For Earth to become another Venus, we would first have to burn 10 times the quantity of fossil fuels our planet now holds. The slogan "Save the Earth," laudable as the sentiment may be, misses the point and fails to convey the far more disturbing truth: Earth cares not a whit for us. It endured billions of years before our emergence and will continue to endure billions of years after the last *Homo sapiens* takes a final breath. As Dr. Ian Malcolm says in the 1993 film *Jurassic Park,* "Life finds a way." But "life" may not necessarily include humanity.

EARTH-MOON SYSTEM

"We set out to explore the Moon and instead discovered the Earth." —William Anders, Apollo 8 astronaut (photographer of "Earthrise")

When Neil Armstrong became the first person to put his feet on the surface of Earth's Moon, he carried in his left pocket a narrow,

extendable metal spoon attached to a small bag. His first and most urgent scientific directive, before planting a flag or taking a single hop, was to scoop up a bit of moondust and shove it into his pocket. The reason? If the mission went awry and he and his fellow Moonwalker, Buzz Aldrin, had to hightail it out of there, scientists on Earth would still have something to study. Amid the emphasis on political ramifications and ticker-tape parades, we often forget that the Moon landing was primarily a scientific mission.

That bag of dust made it back to Earth, along with 48 pounds of lunar pebbles and core samples. This Moon matter, along with another 800 pounds of it returned by the five subsequent Apollo missions, would rewrite the origin story not only of the Moon but also of the entire solar system—and maybe even the origin story of life itself. Before humans traveled out beyond our own planet's atmosphere, little was known about the other worlds of our solar system or how they got to be where and what they are.

Before the Apollo missions brought batches of lunar rocks to Earth for thorough study, scientists held three hypotheses about the formation of the Moon:

1. The Moon formed nearby and at the same time as Earth.
2. The Moon formed elsewhere but was captured by Earth's gravity when it wandered into Earth's path.
3. Earth once spun so fast that parts of it flung off to form the Moon.

Analysis of the Moon rocks, intended to corroborate one of these leading ideas, instead inspired an entirely new concept: the giant-impact hypothesis.

The story goes like this: About 100 million years after the Sun formed, many young planets—protoplanets—caromed around

The violent collision of a Mars-size object and a young Earth may have dislodged the debris that eventually formed the Moon.

COSMIC CONUNDRUM

LIFE WITHOUT THE MOON

Whatever scenario formed the Moon, the fact remains that without it, life on Earth would be either impossible or unimaginable. Earth rotates with a 23.5-degree tilt. This tilt is what causes our seasons. When the Northern Hemisphere tips toward the Sun, it experiences summer; when it tips away, it experiences winter. The same is true for the Southern Hemisphere. The Equator, eternally caught between the two, experiences no seasons. Absent the Moon's constant pull of gravity, which stabilizes our tilt, Earth's axis would bob dramatically, shifting rapidly from seasonlessness to full-blown ice ages. In such an unstable environment, life-forms would have had great difficulty developing, let alone surviving for any length of time.

the early solar system, smashing into one another in the ongoing contest of cosmic billiards that would settle all celestial bodies into their current positions. One Mars-size protoplanet sideswiped the nearly fully formed Earth, creating a ring of debris that birthed our Moon and altering the fate of Earth. Scientists have even honored this hypothetical object with a name from Greek religion: Theia, after the mother of Selene, goddess of the Moon.

Picture this: an illuminated arc across the sky, with newborn rings of flotsam and jetsam whirling overhead. Within those rings of debris, the force of gravity doing its duty makes large pieces larger as they swallow smaller pieces. The ring material rapidly amalgamates into a single huge object—our Moon. Mounting pressure and friction create so much energy that the Moon is a ball of red-hot magma. The new Moon is about 20 times closer to Earth than it is now, rendering it 400 times brighter.

The fate of Theia itself remains a mystery. Most likely, if it existed at all, it disintegrated upon impact, contributing its mass and composition to both the young Earth and the new Moon. And although the tale of a giant impact has been continually contested and modified, including by ongoing studies of Apollo rocks, it remains the leading hypothesis for the Moon's formation.

The Apollo samples show a Moon whose composition is strikingly similar to Earth's crust: an important clue pointing toward the sideswipe hypothesis. Had the Moon been a vagabond object captured by Earth's gravity, the variance would be far greater. The samples also suggest that the lunar surface was once a magma ocean, with heavier minerals sinking to the bottom and lighter ones drifting to the top. That's right, rocks can float—but only within a denser ocean. The most convincing evidence for a giant impact, however, comes from the seismological detection instruments placed by the Apollo astronauts. Data from those devices revealed a small lunar core containing very little iron—a strange discovery for such a sizable object.

Considering its location in the solar system, the Moon, for a moon, is rather large compared with its host planet. It's half the width of Mars, itself host to two asteroid-size, potato-shaped moons. In other words, if the Moon were native born rather than hewn from Earth's crust, it should contain a core of heavy elements, as do all the other large spherical objects in the solar system, which were drawn from the ingredients of our original solar nebula.

> Considering its location in the solar system, the Moon, for a moon, is rather large compared with its host planet. It's half the width of Mars, itself host to two asteroid-size, potato-shaped moons.

This doesn't mean that everybody's doubts have been erased. The giant-impact hypothesis fails to reconcile several geochemical inconsistencies, and Apollo sampled a frustratingly tiny portion of the Moon. So the origin story remains largely hypothetical, at least until humans return for deeper sampling and further study.

THE TIDAL FORCE

When Isaac Newton realized the falling apple was governed by the same law that kept our Moon in orbit, he wrote a recipe to compute the force of gravity between any two objects in the universe. Applying Newton's equation, you can show that the closer two objects are to each other, the greater the force of gravity. As you stand on Earth, for example, Earth's gravity is slightly stronger at your feet than at your head. The difference is smaller than small, so don't blame your light-headedness on it. Earth pulls on your feet with a force only 0.00006 percent stronger than at your head.

All objects feel this simple difference in gravity, officially known as the tidal force, as the gravity of all other objects in the universe pulls them. Tidal forces are the direct cause of a diverse array of cosmic phenomena that otherwise seem to have nothing to do with one another.

A tidal force is strongly dependent on distance—a mild increase in distance between two objects can make a large difference in its strength. For example, if the Moon were just twice its current distance from us, then its tidal force on Earth would be eight times weaker. At its current average distance of 240,000 miles from Earth, the Moon manages to create considerable atmospheric, oceanic, and crustal tides by attracting the part of Earth nearest the Moon more strongly than the part of Earth that is farthest. The Sun, on the other hand, is so far away that in spite of its strong gravity, its tidal force on Earth amounts to half that of the Moon.

The 45-mile-deep outermost layer of the Moon is thicker than Earth's crust. Beneath it is a 620-mile-thick mantle and a small solid, iron-rich core.

The most visible consequences are the tides, as the oceans stretch toward the Moon. Meanwhile, as the solid Earth continues to rotate, the continental shelves keep pushing the quintillion tons of seawater in Earth's oceans forward. This "force war" creates an oceanic bulge, which is always found slightly ahead of the Moon's location in its monthly orbit. Rotating within the bulge, Earth suffers enormous friction as the ocean water sloshes against the continental shelves and shores. The consequence? With the passage of time, Earth rotates ever more slowly. Today, the days are getting longer at a net rate of about two milliseconds per day per century. That doesn't sound like much, but at that rate, full seconds add up fast. It means after one century, every day is two milliseconds faster. After two centuries, days are four milliseconds faster, and so forth. Since 1972, we have been officially adjusting our daily time reckoning with leap seconds that, by calendar decree, are added every few years—as necessary—at the end of June or December.

Go back in time and imagine the tidal situation right when the Moon first formed. At about 20 times closer to Earth than it is today, the Moon's tidal force would have been 8,000 times stronger than now. Some estimates suggest that Earth's much faster rotation rate back then made for a four- or six-hour day instead of our 24. We've been slowing down ever since.

> Imagine the tidal situation right when the Moon first formed. At about 20 times closer to Earth than it is today, the Moon's tidal force would have been 8,000 times stronger than now.

The best evidence for Earth's slowing rotation comes from detailed records of total solar eclipses that date back many centuries. If Earth's rotation rate was faster in the past, then a total solar eclipse as seen on Earth's surface would miss the expected spot that we would otherwise expect it to be at our current rotation rate.

Historical records show precisely that—the earliest recorded eclipses were offset along Earth's surface by somewhere around 6,000 miles west of where they may have been witnessed on today's more sluggishly spinning planet.

Meanwhile, Earth's bulging gravity field, positioned slightly ahead of the Moon in its orbit, acts as an energy pump, slowly rocking the Moon into an ever larger orbit. Need evidence of this? In 1969, when Neil Armstrong and Buzz Aldrin visited the Moon's Sea of Tranquility, among the things they left behind was a mirror-studded panel designed to reflect light in exactly the same direction from which it arrived. Starting shortly after the Moon landing, initially at the McDonald Observatory in Texas and continuing today at observatories in France, Germany, and Italy, high-powered lasers on Earth are beamed up to the Moon, and the return signal is carefully timed.

Knowing the speed of light means we can compute the Moon's distance with unprecedented accuracy. Enlightened by our multidecade baseline of measurements from multiple reflectors, we now know that the Moon is spiraling away from us at a rate of about one and a half inches a year, just as predicted by tidal theory. Earth's rotation will continue to slow down, and the Moon will continue to spiral away until one day on Earth exactly equals one lunar month. By then, one Earth rotation will last more than a thousand hours, which would require adding four million leap seconds a day. No need to panic just yet, though. We've still got more than a trillion years to think about it.

Meanwhile, Earth's tidal force on the Moon finished its work long ago: The Moon's rotation has slowed to exactly equal its period of revolution around Earth. Whenever that happens to any orbiting object, it will always show the same face to the body it orbits—it becomes tidally locked. This is why, as seen from Earth, the Moon has a permanent near side and far side—and when viewed from the near side of the Moon, Earth never sets. During a full lunar month, however, all sides of the Moon receive sunlight. So, contrary to common parlance, folklore, and the title of Pink Floyd's best-selling 1973 rock album, there is not now, nor was there ever, a "dark side" of the Moon.

Remember also that Earth still has a lot of slowing down to do. When its rotation slows until it exactly matches the orbital period of the Moon, then Earth will no longer be rotating within its oceanic tidal bulge, and the Earth-Moon system will have achieved a double tidal lock. (Sounds like an undiscovered wrestling hold.) It so happens that double tidal locks are energetically favorable—somewhat like a ball coming to rest at the bottom of a hill. They're common in the universe in closely orbiting double-star systems. We even have a couple in our own solar system. Earth-Moon and Pluto-Charon are orbiting pairs in which the satellite is nearby and relatively large compared with the host, a configuration that leads to strong tidal forces. Whereas Earth has tidally locked the Moon, Pluto and Charon have tidally locked each other.

Sometimes people wonder whether the Moon's tidal forces can affect human behavior. The answer is yes, provided you have a very, very big head. If your brain were, say, 8,000 miles in diameter (the average diameter of Earth), then the Moon's tidal forces would indeed give you a measurably oblong cranium and impart untold consequences on your mental faculties. For normal *Homo sapiens,* however, the difference in the effect of lunar gravity from one side of the head to the other is immeasurably small, squishing your head by one thousandth of one millimeter. The weight of your own 10-pound head imparts a force significantly greater than the Moon's tidal pull on it—a fact not shared by those who write about werewolves and other Moon-caused dysfunctional behavior.

MARS

Though Venus helped uproot the idea of geocentrism, at least among astronomers, Johannes Kepler's analysis of Mars overturned a concept even more deeply engrained in the worldview of the time: circular motion.

Fundamental to the ancient and medieval views of the solar system was an obsession with spheres and circles, considered to be the most perfect natural shapes. To account for the retrograde motions of the celestial bodies, Ptolemy assigned the backdrop of stars and the wandering planets to separate crystal orbs. Each celestial body nested inside its own orb made of an invisible substance called the aether. Everything moved in a perfect circle around Earth.

When Copernicus proposed his simplified heliocentric model, he maintained the concept of circular motion. So ingrained in astronomical thinking were spheres and circles that relinquishing the idea of circular paths was nearly as inconceivable as relinquishing Earth's position in the center of the universe. But as methods of measurement advanced and the geocentric model began to falter, the assumption of circular motion would fall alongside it.

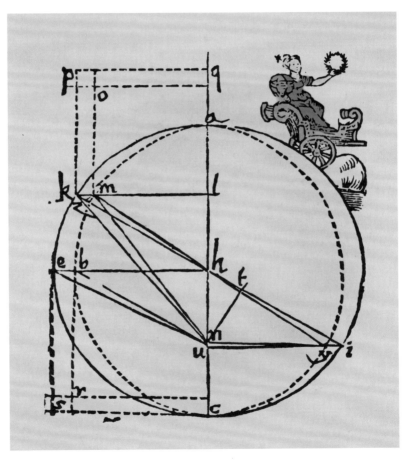

German astronomer Johannes Kepler (1571–1630)
devised the three fundamental laws of planetary motion,
conceptualized here with the orbit of Mars.

In 1601, Johannes Kepler moved to Prague to work for Tycho Brahe, the world-renowned court astronomer of the Holy Roman Empire. Brahe was dedicated to careful observation and measurement. With the help of his state-of-the-art "sighting tubes" and increasingly good data on planetary motions, he encountered a confounding problem: Mars. The Ptolemaic geocentric model of the universe had worked well enough to permit people to think the universe was quite orderly. But with the more precise measurements Brahe's instruments afforded, inexplicable incongruities emerged.

Brahe contemplated the Copernican model, the Ptolemaic model, and even his own combination of the two. But no matter what he tried, Mars never showed up where it was predicted to be. So Brahe off-loaded the Mars problem to his new assistant while he busied himself with the solar system as a whole.

To resolve the problem of the recalcitrant planet, Kepler had to break away from the most basic principle that underpinned the presumed structure and motion of the universe: circular orbits. Only an orbit in the shape of a flattened circle—an ellipse—could explain Mars's anomalous movement. Unlike a circle, which has a singular, central focus, an ellipse has two foci. In the solar system, the massive Sun sits at one focus and nothing sits at the other.

Kepler published his revelations in 1609 in *Astronomia Nova (New Astronomy)*, where he also recorded the first two of his laws of planetary motion, followed 10 years later by his third:

1. The law of orbits: All planets move in elliptical orbits around the Sun.
2. The law of areas: A line that connects a planet to the Sun sweeps through equal areas in equal time. In other words, the closer a planet is to the Sun, the faster it moves.

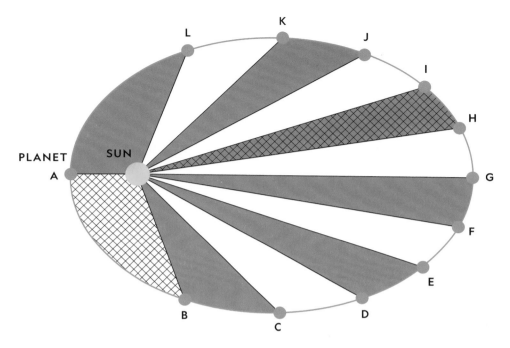

In accordance with Kepler's second law, all these
triangles equal one another in area.

3. The law of periods: The duration and size of a planet's orbit
   are mathematically correlated. The larger the orbit, the longer
   the planet takes to complete it.

These three laws of planetary motion helped lay the groundwork
for Isaac Newton's discovery, decades later, of his own three laws
of motion and gravity.

But most major thinkers of the time, including Galileo and
Descartes, remained so attached to the idea of perfectly circular
celestial motion, whether for metaphysical, philosophical, or
religious reasons, that Kepler's radical concepts were slow to
take hold. Gradually, however, the Copernican system and
Kepler's laws cemented themselves, spurring new discoveries
and culminating, in the 18th century, in what the philosopher
Thomas Paine called the age of reason.

The giant Valles Marineris canyon system on Mars stretches almost 2,000 miles, dwarfing the Grand Canyon.

IMAGINED MARTIANS

In our own day, more than 400 years after Kepler cracked the Mars quandary, the red planet remains the most studied celestial body aside from Earth and its Moon. Before an armada of flybys, orbiters, and rovers was unleashed to scour it up close, humans maintained a hope of discovering Mars to be an Earthlike world, perhaps populated with intelligent extraterrestrial life.

In 1695, during his final days, the Dutch mathematician Christiaan Huygens finished writing the first scientific speculation on this subject. Published posthumously, it was titled

Cosmotheoros: Or, Conjectures Concerning the Inhabitants of the Planets.

Decades earlier, using a better telescope than Galileo's, Huygens first identified Saturn's rings and its moon Titan. He also accurately estimated a Mars day to last about as long as an Earth day and drew the first map of the red planet's huge central plain, Syrtis Major. For Huygens and many other thinkers of his time, the idea that extraterrestrial life might dwell upon these ever emerging worlds was obvious—and exciting. But Huygens took it a step further when he argued that the existence of life beyond Earth was not only compatible with Christian Scripture but also an inevitable part of its teachings, writing:

> Now should we allow the Planets nothing but vast Deserts, lifeless and inanimate Stocks and Stones, and deprive them of all those Creatures that more plainly speak their Divine Architect, we should sink them below the Earth in Beauty and Dignity; a thing that no Reason will permit . . .

By the late 19th century, telescopes were sufficiently advanced to make out some of Mars's topographical features. When the Italian astronomer Giovanni Schiaparelli viewed the planet during opposition—Mars's closest approach to Earth, occurring every 26 months—he was surprised to see networks of straight channels cutting across the surface. Upon declaring his discovery, he used the Italian word *canali* to describe what he saw. If you don't know Italian but think you do, you might mistakenly translate that word as "canal"—a word widely heard in Schiaparelli's day, as the Suez Canal, that marvel of human engineering, had recently been completed.

It's one thing for Mars to have channels. It's quite another for it to have canals.

As it is wont to do, human imagination leaped to conclusions. The American astronomer Percival Lowell, who could have benefited from more Italian lessons, proposed the canal hypothesis, with intelligent Martians digging waterways that crisscrossed the surface of the planet so as to transport their precious, winnowing water supply from the planet's icy poles to the equatorial regions where it was needed. Inspired in part by this concept, legendary science fiction writer H. G. Wells wrote one of the most famous tales of all time, *The War of the Worlds,* which told of an invasion of Earth by hostile, intelligent Martians in search of a new home after their own planet becomes dry and depleted.

Despite Lowell's insistence—and his many-year mapping of the lines he claimed to see—most astronomers failed to observe the supposed canals. Even the most careful and rigorous viewings of Mars at opposition, using the most powerful telescopes available, returned no evidence of their existence. Yet Lowell's lines found favor among the general public and science fiction fans and even appeared on the planning maps for NASA's Mariner 4 mission to Mars.

By the end of the 1960s, however, later Mariner missions and advanced telescope technology captured enough of the Martian surface to kill the idea of canals. It became apparent that they were simply optical illusions, caused chiefly by the human tendency to see what we wish to see, rather than what is.

> Since Lowell's observations predated astronomical photography, he reported on what he believed he saw, not on what was actually there. His incorrect yet adamant claims disintegrated into delusions when met with empirical evidence.

Since Lowell's observations predated astronomical photography, he reported on what he believed he saw, not on what was

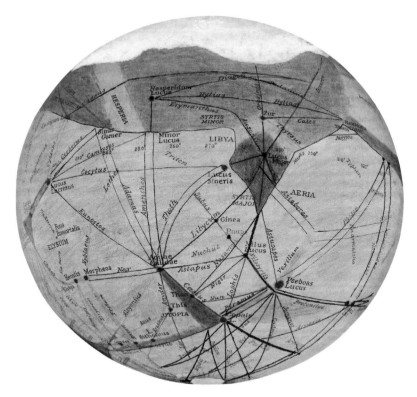

Percival Lowell's color drawing (1905) shows the canals
he believed traversed the surface of Mars.

actually there. His incorrect yet adamant claims disintegrated
into delusions when met with empirical evidence—a humbling
lesson as we continue our exploration of the solar system and
beyond.

EXPLORING MARS

The very first close-up images of Mars, returned by Mariner 4 in
1965, revealed a cratered surface. The probe also returned information on the planet's thin atmosphere, freezing temperatures,
and weak magnetic field, dashing any residual hopes of complex
life currently flourishing there. Even so, Mars remains the most
Earthlike planet we know of. It may even have looked like Earth

TRIPLE POINT OF WATER

I f you've ever prepared a meal at a high elevation, you know that you need to adjust cooking times because water boils at a lower temperature under the lower atmospheric pressures found at higher altitudes than it does at sea level. So, when you're making tea on a mountaintop, the boiling water isn't as hot as you're accustomed to, and you'll need to steep your tea bag a bit longer. If you keep ascending, the pressure continues to lower, and the boiling point of water continues to drop. Eventually, you'll encounter an atmospheric pressure at which the boiling point is the same as the freezing point. If you ever ascend that high into the atmosphere, you'll be able to simultaneously sustain ice, steam, and liquid water in the same pool. The temperature and pressure at which a substance can coexist in these three states of matter is its triple point.

The surface of Mars would be a good place to try that experiment. In some areas you could drop an ice cube into a vat of boiling water and watch the two coexist. But if you tried to brew tea in that Martian water, the tiniest fluctuation in pressure would spontaneously turn your liquid beverage to either solid ice or vapor, offering an unpleasant roulette of a sipping experience.

sometime in its ancient past—a fertile world of lakes, glaciers, perhaps even vegetation.

A decade after Mariner 4 returned its first fuzzy Martian images, the Viking mission—two orbiters, each with its own lander—revealed the planet's physiography and dynamic history in detail. Valleys and grooves carved into the red planet's surface suggested a vast, vanished network of rivers, lakes, and floodplains.

As Mars cooled over the eons, the core solidified and the planet could no longer sustain a protective magnetic field. As far as we know, only a churning molten metal core makes that possible. A magnetic field shields a planet's surface from the solar wind's high-energy particles, which would otherwise wreak havoc. But Mars has no such protection. Most of its atmosphere and nearly all the liquid surface water evaporated eons ago, leaving behind only traces. Any water that does still exist on the planet is either trapped in the polar ice caps or buried deep underground.

Today, Mars is a frigid tundra, with an average surface temperature of minus 80°F—cold enough to make frozen Siberia seem positively balmy. Here on Earth, you don't see people scrambling to settle the North or South Pole. Nonetheless, if humans are ever to become a multiplanetary species, the most logical migration destination would be Mars.

TERRAFORMING MARS

Mars comes with an Earth 2.0 starter pack: a bit of atmosphere, some gravity, and (we think) reservoirs of water ice beneath the surface, lurking as permafrost. Just add heat. Some assembly required.

Unless humans want to live inside, and at the mercy of, artificial habitats, we'll want to figure out how to live and breathe on the surface by adapting the environment into one suitable for our frail human bodies. In other words, we'll want to terraform it. Elon Musk's dream of nuking Mars into a hospitable temperature zone is one drastic approach to terraforming. While that might sound like something a comic-book supervillain might think of, it could work—in principle.

We've known about Martian ice caps for quite some time, and our Mars rovers and orbiters continue to detect H_2O underground. So, yes, detonate enough bombs near the poles, where

most of Mars's ice lives, and much of the ice would melt and release its rich stores of carbon dioxide and water vapor—greenhouse gases that could cloak the planet and trap heat. This method has a few problems, though:

1. It could trigger another cold war.
We would need to detonate thousands of nuclear weapons to create sufficient heat. While the logistics of transporting that many bombs through space seems implausible all on its own, somebody would have to be in charge of creating and overseeing their creation—and if history has anything to teach us, the fears surrounding anybody's creation of nuclear weapons can be as detrimental to humanity as their actual detonation.

What might it look like? This time-lapse simulation
envisions a terraformed Mars.

2. It could trigger a nuclear winter.

Mars is dusty. If the bombs kicked up enough debris to
block sunlight, the temperature of the planet would drop
dramatically, rather than rise. In fact, that's what killed
most of the dinosaurs and three-quarters of other life-
forms on Earth some 66 million years ago when an asteroid
hit Mexico's Yucatán Peninsula. So much ash and dust were
ejected into the atmosphere that the skies darkened and
photosynthesis ground to a halt, knocking out the base of
the food chain and sending waves of extinctions across the
world, working their way up through the tree of life.

3. It brings into question the Prime Directive.

Every *Star Trek* voyager was subject to the Prime Directive, a bit of space law that explicitly forbade meddling with alien planets or life. Are we? NASA's Office of Planetary Protection works to implement policies and offer advice on how to prevent extraterrestrial life from contaminating Earth and to prevent Earth life from contaminating any other place in the solar system that may harbor life. We're still not sure whether Mars hosts (or ever hosted) life of its own. If it does, what are the ethical implications of destroying their world? Life on Mars, if it does exist, might be tantamount to pond scum. Does pond scum have a right to live?

Here's a gentler, albeit equally bizarre alternative that just might work: giant space mirrors. American engineer Robert Zubrin and NASA planetary scientist Christopher McKay have proposed that a gigantic mirror placed in orbit around Mars, and positioned just so, could reflect and redirect sunlight toward the Martian poles and melt the ice. As long as the mirror stayed where it was supposed to, the method would pose far less threat to future human settlers. Sounds good. But our old adversary, the rocket equation, will likely put a damper on this plan. A mirror of the size proposed—almost 80 miles wide and weighing a couple hundred thousand tons—would prove an impossible payload to launch from Earth. Such a project would require space construction, using materials mined in space—much cheaper than hauling stuff up from Earth.

An inescapable problem with any terraforming scheme is that Mars simply does not contain enough CO_2 to trigger global warming. In fact, some scientists think that to trigger any warming at all, we would need to release more CO_2 on Mars than humans on Earth have ever released.

In any case, if humanity ever develops enough geoengineering know-how to terraform Mars as our escape plan after we trash Earth, then we should certainly be able to use that intelligence to make Earth livable again and save ourselves from requiring a Planet B in the first place.

THE ASTEROID BELT

Between the orbits of Mars and Jupiter lies a ring of debris, asteroids, and one dwarf planet, separating the four rocky inner planets from the four gassy outer giants. The dividing line is made of leftovers from the game of cosmic billiards that played out four and a half billion years ago.

Long before any of these were discovered, Johannes Kepler reasoned that the space between Mars and Jupiter was far too empty. A planet must exist between the two, he thought. Other astronomers agreed, and for two centuries, the hunt was on for the missing planet.

On the very first day of 1801, an Italian priest and astronomer named Giuseppe Piazzi accidentally discovered a celestial anomaly while cataloguing star positions—a "new star" that was "a little faint and colored as Jupiter," as he wrote in his journal. When the light moved against the backdrop of the rest of the stars, he knew he'd discovered something new within our solar system. At first, he thought the moving light was a comet, and he alerted the astronomical world to a discovery that he called Ceres Ferdinandea. Eventually, improved calculations of its orbit pointed to something much bigger than a comet—something we in the 21st century now classify as a dwarf planet. Soon more "planets" were discovered in the same region of the sky, and for a while, our solar system boasted 11 planets: Mercury, Venus, Earth, Mars, Vesta, Juno, Ceres, Pallas, Jupiter, Saturn, and Uranus.

British astronomer William Herschel noticed that these newcomers all appeared only as points of light in the telescope, just as stars do. Whatever they were, he realized, they must be much smaller than bona fide planets. And so Herschel proposed they be called asteroids—"star-like," from the Greek *aster,* meaning "star." By the mid-19th century, so many more purported planets were discovered that astronomers were ready to adopt Herschel's classification. Today we recognize that the asteroid belt—the region between Mars and Jupiter—is home to hundreds of thousands of objects. One is even named 13123 Tyson.

POTENTIALLY HAZARDOUS ASTEROIDS AND COMETS

Imagine this: A massive asteroid is heading our way, and the laws of orbital mechanics eliminate any hope of a miss. It's big, maybe half a mile across. And its closing speed is tens of thousands of miles an hour. How would we handle this threat?

Several high-budget films have explored this scenario, including the star-studded *Don't Look Up* (2021), which satirized contemporary antiscience thinking. Both the films *Deep Impact* (1998) and *Armageddon* (1998) invoke science and engineering attempts to destroy the deadly interloper.

The scripting and scenes of *Armageddon* itself surely violate more laws of physics per minute than any other film ever made. But that makes it all the more interesting to analyze. *Armageddon*'s solution was to train a crew of oil rig workers to become astronauts. They would then drill deep into the incoming asteroid and plant a nuclear bomb that would crack the asteroid in half and send each piece onto a path away from Earth. Holding aside the question of whether it is easier to train astronauts to drill an asteroid than to train oil drillers to become astronauts,

SMASHING THROUGH THE ASTEROID BELT

We all know the scene: A spaceship hurtles through space, perhaps pursued by oncoming enemies. Suddenly a dense thicket of huge, tumbling asteroids appears, threatening to bash their ship and kill everybody on board. In the 1980 film *Star Wars: Episode V—The Empire Strikes Back,* Han Solo and his crew find themselves in just such a predicament with a pack of imperial TIE fighters on their tail. "You're not actually going into an asteroid field," says an alarmed Princess Leia. Han Solo replies, "They'd be crazy to follow us, wouldn't they?" C3PO helpfully chimes in, "Sir, the possibility of successfully navigating an asteroid field is approximately 3,720 to one." Of course, the congested swarm of careening rocks turns out to be no match for Solo's piloting skills. "Never tell me the odds," Solo retorts. Classic.

The ubiquitous trope tends to lure people into believing that it poses a common and extreme threat. But careful observations of our own solar system tell us that a lingering asteroid minefield as depicted in movies doesn't exist. If we collected all the asteroids and the lone dwarf planet from the asteroid belt into one big heap, all that matter would add up to about 3 percent of the total mass of Earth's Moon. If you picture a small chunk of Moon spread across the sweeping expanse occupied by the asteroid belt, you can understand how inane this trope really is. Any probe sent through the belt is not only unlikely to collide with an asteroid—it's unlikely to come within a quarter million miles of one. Plus, a sparsely populated asteroid belt is not merely a one-off feature of our own solar system; it's the cosmic trend. Any field of objects as crowded as the ones portrayed in Hollywood would have coalesced into one larger object, thanks to gravity. If *Star Wars* took place in reality, then when Han Solo and his team arrived at the "asteroid field," they wouldn't even realize it.

the asteroid would not crack in half. No, it would explode into countless unpredictable pieces, many of which would head toward Earth even faster.

Furthermore, a group of physics students at the University of Leicester in the United Kingdom calculated that the *Armageddon* solution would require a bomb approximately a billion times more powerful than the most powerful bomb ever detonated on Earth. As we said, the film played loose with physics.

The best solutions are far less dramatic. A mild nudge in a different direction would suffice to knock an asteroid off course and out of our path. That nudge could come from a variety of tactics:

1. Spray paint it.
Ridiculous though this may sound, spray paint could steer an asteroid away from Earth. All you need to do is paint half the thing white. The lighter side would then reflect more of the Sun's energy than the darker side, causing a momentum imbalance. Enough of this pressure difference over enough time would cause the asteroid to veer off course. But what if we can't get a giant paintball gun to it in time?

2. Slow it down.
We could position a series of objects to act as stumbling blocks in the asteroid's path. Each impact would absorb some kinetic energy and slightly slow down the rock as Earth moves out of harm's way via its usual orbit. But what if one misses its target?

3. Thwack it with a space probe.
Launched in November 2021, NASA's DART (Double Asteroid Redirection Test), a spacecraft the size of a car, was conceived to test whether an object smacked into the side

An artistic rendering of the debris-filled asteroid belt around the star Vega of an oncoming asteroid might be enough to alter its course. Ten months later, Earthlings seven million miles away watched the livestream feed of the DART spacecraft as it approached its target, Dimorphos, a moonlet of the asteroid Didymos. A gray, craggy world emerged, growing larger and closer until finally the feed went blank. Direct impact.

Could such a comparatively tiny object really alter an asteroid's course? It's like a couple of linebackers running headfirst at the Great Pyramid of Giza. But, hurtling through the vacuum of space at 15,000 miles an hour, even a small mass can deal serious damage. Days later, a 6,000-mile tail of dust and rubble from the collision streaked across our sky as a final salutation from the pulverized DART spacecraft. Scientists confirmed an orbital shift of the asteroid—another cosmic first for humanity.

4. Nuke it.
Unlike in *Armageddon,* a nuclear bomb would be positioned just above the asteroid's surface. Its detonation would vaporize a portion of the surface and thus slightly alter the object's trajectory.

5. Wield the force of gravity against it.
Because everything in the universe exerts gravity, we could launch a spacecraft to come close to, but not touch, the oncoming asteroid. The two objects would be drawn toward each other, but our spacecraft would fire a series of station-keeping jets—just enough to keep the asteroid at a slight distance and gently, over time, tug it off its original course.

In every case, the sooner we act, the better the chance of a successful deflection. The farther away the oncoming object, the

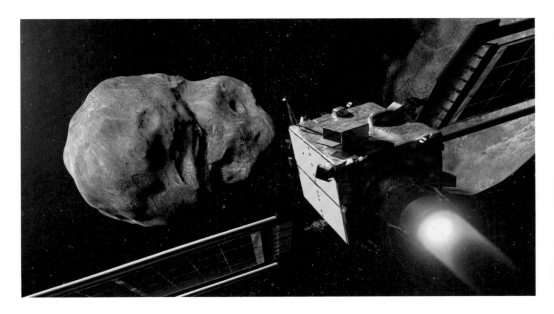

An illustration of NASA's Double Asteroid Redirection Test (DART) spacecraft prior to impact at the Didymos binary asteroid system

smaller the deviation needed to send it off course. The closer it gets, the more dramatic the deflection must be.

Before his death in 2018, astrophysicist Stephen Hawking warned that the greatest threat to humanity was an asteroid impact. Two years earlier, NASA had set up a department dedicated solely to monitoring and preventing this scenario: the Planetary Defense Coordination Office. To the best of our knowledge, the large dinosaurs didn't operate such a program; if they had, they'd likely still be around. After all, they roamed Earth for nearly 200 million years before the asteroid hit Chicxulub, and only 66 million years have passed since then. By the way, that means you live about as close in time to a *T. rex* as the first *T. rex* did to the last *Stegosaurus*. That's how stupefyingly long the various dinosaur species dominated Earth.

Space agencies around the world routinely find and track potentially hazardous asteroids (PHAs) and comets. This scary subset of near-Earth objects currently numbers about 2,300. Although almost none of them poses a threat in the coming century, it's possible or even likely that a few lurk in the vast expanses of space, as yet unaccounted for. Should the time ever come to act, we'd better be prepared—and be listening to the scientists.

THE GAS GIANTS

Beyond the asteroid belt is the realm of gas giants and frozen worlds, made of volatile materials. There reign Jupiter and Saturn—and, farther out still, the ice giants Uranus and Neptune.

Like so many stories of our solar neighborhood, the origin of the gaseous orbs remains contested. Once upon a time, scientists presumed that all the planets in our solar system formed exactly where they are found today, slowly amalgamating matter dispersed throughout their orbits into a series of separate large bodies. But the Moon rocks and craters that dot the rocky inner planets point to a much more tumultuous beginning.

One of the leading hypotheses, called the Nice model—pronounced like "niece" and named after the coastal city in the French Riviera, where the idea was hatched—is inspired in part by how many known exoplanets are Jupiter-size and orbit very close to their host star. It proposes that the giant outer planets were once much closer to the Sun. Surrounding the Sun was also a large ring of icy planetesimals and other chunks of leftover debris from the solar system's initial formation. The giant planets tugged on some of the outer debris and sent it careening toward the inner solar system, resulting in what's called the late heavy bombardment period of the solar system. In response, the giants' own orbits slowly migrated farther and farther away from the Sun. Multitudinous small interactions over time, along with the gravitational interplay among these planets, would have reshuffled everything into their current positions. If the hypothesis holds true, all that bombardment and reshuffling would have also delivered water ice and the building blocks of life to the inner planets.

JUPITER

Let us ponder the existence of a place that is not only 1,300 times larger than Earth but also twice as massive as all the other planets combined. That place is the planet Jupiter, named for the king of the Roman gods. When Galileo first observed Jupiter through his telescope, he was shocked to find four clearly distinguishable moons in orbit around it, showing that some celestial spheres contentedly orbit bodies other than Earth. To honor their discoverer, Io, Europa, Ganymede, and Callisto are collectively referred to as the Galilean moons.

> Let us ponder the existence of a place that is not only 1,300 times larger than Earth but also twice as massive as all the other planets combined.

In many ways, Jupiter is a protector of the inner solar system and especially of Earth—even if, in its early days, it did send down an apocalyptic hailstorm of icy debris. Now its massive gravitational pull either consumes many small objects that might be making their way toward the Sun or flings them out on a divergent course.

A SPOT OF RED

Jupiter is a playground of atmospheric dynamics that intensifies all rotationally induced cloud and weather patterns. In a

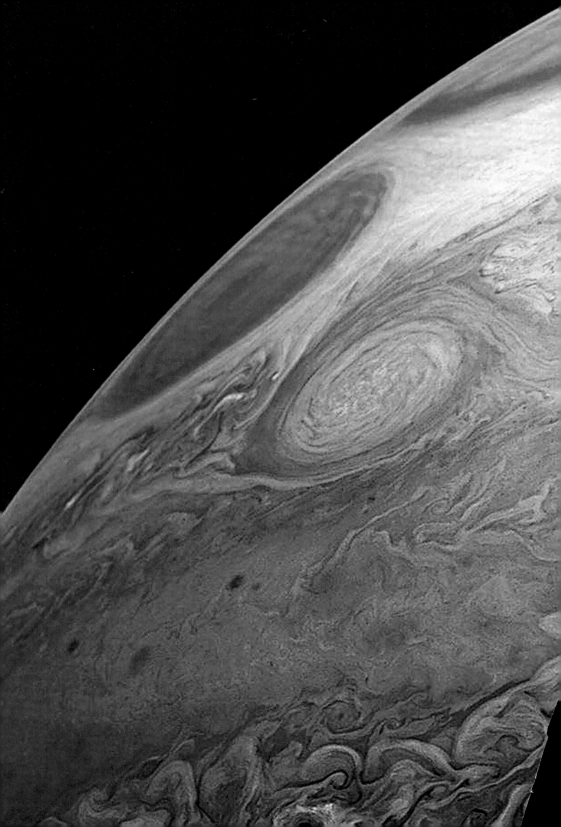

supremely striking display of the Coriolis force, Jupiter endures the largest, most energetic, longest-lived storm in our solar system: an anticyclone that looks like a big reddish blotch in the planet's upper atmosphere. We call it Jupiter's Great Red Spot. Possibly discovered in 1664 by the English physicist Robert Hooke and, much more certainly, the following year by the Italian astronomer Giovanni Cassini, the Spot is still stormy, three and a half centuries later.

The Great Red Spot, by the way, is bigger than Earth and as much as 300 miles deep, although its size and shape have varied over time. It lives in Jupiter's southern hemisphere and rotates counterclockwise, which immediately tells us it's a high-pressure system. Its coloration, ranging from orangish red to a barely visible pale cream, is generally attributed to varying concentrations of phosphorus and sulfur compounds, with a touch of ammonia. Close-up images from the Voyager flyby missions of the late 1970s revealed a maelstrom of colorful swirls at the interface of the Spot and the surrounding atmosphere. Clearly resolved horizontal belts interlaced with countless smaller cyclones and anticyclones also make Jupiter look like an archaeological cross section of a Big Mac, bun included.

What's driving the Spot's enormous storm? So far, we can only make educated guesses. Jupiter radiates twice as much heat as it receives from the Sun. In addition, enormous thermal reservoirs in Jupiter's interior can drive atmospheric flow patterns. One source of the internal heat would be the radioactive decay of trace elements, while another would be the leftover heat from Jupiter's initial contraction from a protoplanetary cloud to a planet during the early phases of the solar system. The Spot's sustaining energy could also (or instead) be tapped from

In 2018, NASA's Juno spacecraft captured this image of Jupiter's roiling southern hemisphere, where the Great Red Spot swirls beside another massive storm called Oval BA.

elsewhere. On Earth, hurricanes are partially driven by latent heat released into the atmosphere when raindrops condense; the heated air then rises rapidly. Maybe a similar mechanism dominates Jupiter's atmosphere as its gases condense toward its liquid interior.

The Spot has also been observed to dine on smaller turbulent eddies in its vicinity, a cannibalism that would yield yet another source of energy. (After hearing about atmospheric phenomena on Jupiter, NBC's meteorologist Al Roker commented, "I once knew a turbulent Eddie." When asked, "What's he doing now?" Al replied, "5 to 10.")

Many questions about the Great Red Spot persist, including when it formed and how long it will last. We know it's more than 300 years old, but it could be much older. Over the past few decades, scientists have noticed it shrinking and narrowing, from an elongated oval into a more circular shape; they've predicted it will become completely circular by the 2040s. At least one scientist contends it could disappear entirely; others are skeptical. New missions, equipped with technologies capable of withstanding Jupiter's extreme temperatures, pressures, and turbulence, offer the only path to resolving the persistent questions about this humongous Jovian storm.

EXPLORING JUPITER

Clues to the goings-on within Jupiter's deeper cloud layers emerged in 1995, when the spacecraft Galileo sailed past Jupiter, 50 million miles away, and parachuted a 700-pound probe that measured temperature, density, composition, light scattering, radiation flux, and lightning. Plunging through the outer atmosphere at more than a hundred thousand miles an hour, the probe relayed data for nearly an hour, until intense pressure forced its transmitter to fail. Although the average temperature

A FAILED STAR

n volume, Jupiter is larger than several stars beyond our solar system. Like our star, the Sun, Jupiter has retained nearly all its birth gases, leaving it mostly made of hydrogen (about 90 percent), some helium (almost 10 percent), and a smidgen of heavier elements. Unlike our star, however, the pressures at its core fall short of those required for thermonuclear fusion, which is why it's been unfairly dubbed a failed star.

You might think that if Jupiter had just a bit more mass, our solar system would instead be a double solar system, though you wouldn't be alive and reading this book today if that were the case. In fact, Jupiter would need more than 70 times its current mass to instigate fusion at its core. So really, it never failed at being a star, because it was never given the opportunity. Instead, you might want to praise Jupiter for its wild success at being a planet.

of Jupiter doesn't exceed a chilly minus 200°F, it's quite possible that temperatures in the gas giant's interior could exceed 40,000°F—about four times that of Earth's. But we still don't know much about what happens within the deepest, hottest layers, including its core—a strangely borderless, diffuse amalgamation of dense material spanning half the planet's radius. Jupiter probably also boasts the largest ocean in the entire solar system—but it isn't made of water. Instead, one of the rarest elements found on Earth fills the Jovian oceans: liquid metallic hydrogen.

On Earth, we think of hydrogen as a gas. But under high pressure, such as deep within Jupiter, hydrogen liquefies and can act like a metal, with all the same rights and privileges. That's the origin of Jupiter's titanic magnetic field. And at the base of Jupiter's striated atmosphere, drops of helium rain fall into an

SHOEMAKER-LEVY

n July 1994, astrophysicists around the world trained their telescopes on Jupiter, awaiting an impending series of cataclysms. Chunks of a comet named Shoemaker-Levy 9 (SL-9 for short) were slated to smash into the planet, becoming the first collisions ever observed between two celestial bodies beyond Earth.

Discovered by Carolyn and Eugene Shoemaker and David Levy only a year earlier, the comet had already been ripped into pieces ranging from a few feet to a mile across by Jupiter's monster tidal force during an earlier flyby. Reencountering Jupiter in 1994, the dozens of crumbled comet parts punched through the planet's southerly regions at 37 miles a second, leaving sustained scars more vivid than the Great Red Spot on its gaseous surface.

The collision energy of most fragments rivaled Earth's mass extinction event 66 million years ago. So whatever damage they caused, you can be sure Jupiter has no dinosaurs. The SL-9 event showed that extinction-level impacts were still possible in the solar system, and that our fragile planet was indeed still vulnerable. The Shoemakers and Levy were at that time among the very few scientists actively scouring the solar system for asteroids and comets. (For this comet to be named SL-9 means the trio previously discovered eight other comets.) Jupiter's harrowing plight soon impelled others to monitor the solar system for dangerous near-Earth objects—NEOs—lest we face a similar fate again. Planetary defense was born.

ocean of liquid hydrogen—something we guessed even before the 2001 hit song "Drops of Jupiter" by the rock band Train swept through radio stations across America. Missed it? It was a love ballad that also casually mentions Earth's atmosphere, the Moon, shooting stars, constellations, and the Milky Way.

SATURN

The most distant of the seven original wanderers is best known for the spectacular ring system encircling it, first observed by Galileo in 1610. These rings have justifiably earned Saturn the moniker "jewel of the solar system." Perhaps if the ancients had seen its beauty up close, they would have named it Venus rather than Saturn, the Roman god of agriculture.

When Galileo turned his new and improved but still quite crude telescope on Saturn, he was stunned to see two large round objects flanking the planet. Though its irregular shape appeared only as a few hazy blobs, he concluded that the planet was a three-body system with two great moons far larger than Jupiter's. A couple years later, he again looked to Saturn, only to find that the two orbs had vanished from view. According to Roman mythology, the god Saturn had eaten his sons soon after birth. And so, in response to the surprising disappearance, Galileo quipped, "Has Saturn devoured his own children?"

The mysterious curved objects returned in 1616, whereupon Galileo replaced his three-body hypothesis with a new idea: Saturn had arms protruding from the sphere, like two pot handles that, when seen face-on, disappeared against the larger body of the planet. For the next 40 years, perplexed astronomers offered various explanations for Saturn's oddly changing form. But like so many questions aroused by planetary observation, new clarity would come only with the increasing sharpness of telescopic images.

In 1656 Christiaan Huygens, having spent a year observing the pot handles with a much more powerful telescope of his own design, proclaimed, "Saturn is girdled by a thin flat ring, nowhere touching, and inclined to the ecliptic." Twenty years later, Giovanni Cassini detected a slight gap between two sections in the ring. That gap, now called the Cassini division, is about 3,000 miles wide. Cassini also (correctly) proposed that

This panoramic view of Saturn, backlit by the Sun, combines 165 images taken by Cassini's wide-angle camera in 2006.

the rings themselves were a myriad of particles that he called moonlets rather than a solid object—a hypothesis unproved until two centuries later—and he discovered four of Saturn's moons. Today we know that the ring system contains some tiny moons of its own and billions of chunks of water ice, ranged across thousands of bands.

If you could stand on Saturn's equator and look up, you might not know the rings exist at all. We now understand that, when viewed from Earth, the bands disappear and reappear from time to time because they are so thin, that when viewed edge-on, they all but vanish. Even though the main ring system extends outward some 175,000 miles from the planet (and its faint outermost ring millions more miles), the entire system is only 30 feet thick in some places; most areas measure less than a few hundred feet. In other words, the system is about three million times wider than it is thick. For your average tortilla to have these proportions, it would need a diameter of 100 football fields.

EXPLORING SATURN

Like all the outer planets, Saturn was mostly a mystery until the late 1970s and early 1980s, when the Pioneer 11 and Voyager probes began to return up-close images and data. They revealed many intriguing spectacles: surface temperatures of minus 300°F, winds ripping through the equator at more than 1,000 miles an hour, and, capping its north pole, a large, strange hexagon. The data also showed that Saturn has a higher concentration of hydrogen than Jupiter. Turns out, it contains so much hydrogen that it is, on average, less dense than water. In theory, that means the whole planet could float in a basin of water—if you could find one big enough and ignored several other laws of physics.

But even after multiple flybys, the innards of the ring-festooned planet were still a puzzle, because Saturn's thick gaseous surface obfuscated visible light that might otherwise reach our cameras flying past. Our understanding of Saturn ramped up thanks to the Cassini-Huygens mission, named for the two 18th-century astronomers who first charted this stunning world and its many moons in infrared as well as visible light.

From 2004 until its dramatic death plunge into Saturn's atmosphere in 2017, the Cassini spacecraft orbited the planet, darting in and out of its larger rings while observing many of its 80-plus moons. One of its findings was that the blue-gold hexagon is a giant hurricane, not unlike the one within Jupiter's Great Red Spot. Why it's hexagonal rather than, say, pentagonal has been modeled, but is still not fully understood.

Some of the most fascinating discoveries made by the Cassini mission and its 700-pound shellfish-shaped Huygens probe come from Saturn's moons, rather than Saturn itself. Huygens landed on Titan, Saturn's largest moon, which marked the first time anything from Earth landed on anything in the outer solar system. There, it revealed a surprisingly Earthlike topography. Beneath thick clouds, among dunes and canyons, liquid methane rains into flowing rivers while reservoirs fill with the runoff.

On Earth, we know methane as the potent greenhouse gas released as termite and bovine flatulence. It also comes from anaerobically rotting plants and animals. On Titan, the frigid conditions mean that the methane molecules manifest in liquid form instead of as gas. Titan is the only other world we know of that can retain individual bodies of liquid on its surface. That capacity, along with the dribbling methane "rain," constitutes a state of affairs akin to Earth's water cycle—a feature critical to the evolution of life on our planet. Could life of some kind—a kind we do not yet understand—exist in those methane seas? Deep below the thick crust, a vast sea of liquid water and ammonia may engulf an entire layer of Titan. Here might reside life of a kind we could understand.

OCEAN WORLDS AND THE SEARCH FOR LIFE (AS WE KNOW IT)

Looking for life in the solar system implicitly means looking for life as we understand it. This is the curse of the single example. We know one planet and one kind of life in the entire universe. *T. rex, H. sapiens, E. coli,* and every other living organism large and small share a few critical characteristics and, ultimately, a single-celled common ancestor. Sometime around four billion years ago, when our young planet was just beginning to cool down from all the volcanic activity and heavy bombardment, a cell born of carbon and other atoms that define organic chemistry emerged someplace in a world of warm liquid water. This cell consumed energy, underwent chemical reactions, reproduced itself—and life began. From that point on, single-celled life reigned supreme on Earth for more than three billion years. These days, when we Earthlings seek life elsewhere in the solar system, we tend to look for primitive organisms like our ancient microbial ancestor.

A world that hosts a deep global ocean of liquid water should be an ideal place to start. If you ask most people where in the solar system they'd find the most of it, they'd probably say Earth. But though Earth is the only planet in our solar system with a water ocean at its surface, we know of several other worlds with subterranean seas, each containing far more liquid water than is found in all of Earth's oceans.

Jupiter's moon Ganymede—the ninth largest object in the solar system (larger even than the planet Mercury)—may have an underground saltwater ocean nine times deeper than the deepest trench in Earth's Moon. Because of its thin but existent oxygen atmosphere and also its magnetic field, unique among moons, astrobiologists—astrophysicists who study life in the universe—are desperate to know what, or who, might dwell in the nether regions beneath Ganymede's surface. Unfortunately, if a vast ocean does exist there, it's buried under a hundred miles of ice and rock. For comparison, the deepest humans have ever dug into Earth's crust, the Kola Superdeep Borehole in Russia's piece of the Arctic Circle, is a paltry 7.5 miles—and that took 20 years, at which point the project ran out of money and was abandoned.

> Jupiter's moon Ganymede—the ninth largest object in the solar system . . . may have an underground saltwater ocean nine times deeper than the deepest trench in Earth's Moon.

Cryogeysers erupt from the surface of Saturn's moon Enceladus.

It's safe to say, then, that we won't be conducting any sample-collecting missions on Ganymede anytime soon.

Some astrobiologists think they'll have better luck searching on Europa, another of Jupiter's 80 known moons. Europa is a quarter the width of Earth, but it likely harbors a subsurface ocean in which twice the water in Earth's oceans sloshes around. Also, clues in the ice patterns on Europa's surface suggest that the ocean below might be warm. Jupiter's colossal tidal forces, combined with tugs from other moons, cause Europa to bend and stretch.

Have you ever played racquetball? The opening call to warm up the ball by hitting it for a while is precisely this phenomenon. The ball distorts, recovers its shape, and distorts again and again, as all that thwacking pumps energy into it and raises its temperature. Europa is likely experiencing something similar that is warming it up from the inside out.

Yet another moon—this one in orbit around Saturn—hides an ocean of liquid salty water beneath an icy crust. Enceladus shows a smooth, glassy surface that renders it the most reflective object in our solar system. At its south pole, Cassini discovered geysers spraying a salty cocktail hundreds of miles high. Darting into the plumes for a better look, the space probe found organic molecules necessary to sustain life as we know it.

While Enceladus's and Europa's icy crusts are significantly thinner than Ganymede's at about a dozen miles thick, drilling through those depths still presents a huge technological hurdle.

But these thick crusts pose another problem: Sunlight can't penetrate the watery depths, and so any resident organisms can't practice photosynthesis. Fortunately for the ongoing search for life as we know it in the universe, not all life on Earth requires sunlight to survive. In the deepest, darkest reaches of our own planet's ocean floor—where the Sun don't shine—entire ecosystems of curious creatures thrive around towering, hellishly hot pillars called black smokers. As the tectonic plates shift and collide, near-freezing seawater seeps into cracks, mixes with upwelling hot magma, and soars to temperatures of above 700°F. The now superheated water erupts in billowing blackish plumes of metals and minerals. When the jets encounter the surrounding cold water, the metals and minerals precipitate out and solidify, shaping themselves into towers that can rise as fast as a foot a day.

Bacteria and archaea that live near the vents draw their food and energy from chemical reactions with the outgushed minerals, rather than from the Sun. Astrobiologists think that organisms not unlike these formed the first ecosystems on Earth; given this, similar organisms could exist elsewhere in the universe, provided they have access to the necessary organic materials and inhabit a warmish environment. When the Cassini mission zipped through Enceladus's plumes, it picked up indications of hydrothermal activity far below the surface. Combine the possibility of deepwater vents with the evidence already established of a warm salty liquid ocean, rife with organic molecules, and the idea of a world teeming with alien life transcends science fiction.

ICE GIANTS: URANUS AND NEPTUNE

The realm of the ice giants lies not millions, but billions, of miles from Earth. More than a half century has passed since humans left their boot prints on the Moon and inserted spacecraft into Martian and Venusian orbit. Yet the farthest planets of our solar system still await orbiters and landers of their own. Of the hundreds of missions dispatched to explore space, only one probe, Voyager 2—after eight and a half years of traversing space and multiple gravity-assist maneuvers—even caught a passing glimpse of Uranus, and then Neptune three and a half years later. Unlike Jupiter and Saturn, which are made mostly of hydrogen and a bit of helium, Uranus and Neptune are perhaps one-fifth hydrogen and helium. Water ice is their main component.

Despite its brief flyby, Voyager confirmed what scientists had long suspected: The least explored region of our solar system is the most bizarre. Beneath the thick gaseous atmospheres of these two most distant worlds in our solar system, diamonds rain down into slushy oceans of water, methane, and ammonia— or so atmospheric data and carefully constructed models suggest. Their stormy, nearly impenetrable cloud covers keep much of their inner workings a secret.

When Voyager 2 turned its cameras on Uranus, they registered a giant, uniformly colored aqua ball. Lacking the dramatic,

WHAT'S IN A NAME?

Adding to Uranus's list of oddities is the history of how it got its name, which of course has made it the butt of many jokes. Detected in 1781 by William Herschel, it was the first planet ever actually discovered; all the rest had long been known to anyone who looked up. Herschel tried to name the object Georgium Sidus, after King George III—the very same king who impelled Benjamin Franklin, John Hancock, and 54 other notables to sign the Declaration of Independence. The rest of the scientific community objected, and the name never caught on. Since nobody other than Brits were content with an enumeration of planets that would read "Mercury, Venus, Earth, Mars, Jupiter, Saturn, and George," the new planet instead was named Uranus, after the primordial Greek god of the sky. Still, an unusual choice, because planets were traditionally named for Roman gods while their moons were named after Greek characters in the life of the Greek counterparts to those gods. As a concession to the mighty king of England, Uranus's moons are instead named after characters in English literature, primarily from Shakespeare (like Juliet, Puck, and Miranda) plus three from Alexander Pope (Ariel, Umbriel, and Belinda).

swirling storms seen on Jupiter and Saturn, Uranus appeared curiously plain—even boring—at first glance. But it turned out to be an oddity. Most notably, Uranus orbits on its side, with rings hugging it vertically. Unlike every other planet in our solar system, it rotates nearly perpendicular to its orbit—a feature best explained by a hypothetical collision with a huge planet, shortly after its birth, that tipped the whole thing over.

Having zoomed past Uranus, Voyager 2 had a billion more miles to go before reaching the next and last planet, Neptune,

Uranus's moons and faint rings are enhanced for visibility in this 2003 false-color view of the planet from NASA's Hubble Space Telescope.

after which it permanently exited the solar system. To the delight of scientists back on Earth, the probe relayed images of a topographically interesting world. Though chemically almost identical to Uranus, Neptune—as we see in more recent photos—radiates a deeper, more vibrant azure blue than Uranus's pale greenish blue haze, although the cause for the difference is not yet understood. Voyager also revealed thrashing winds of 1,000 miles an hour, the fastest in the solar system.

Striking as the attributes of Uranus and Neptune may seem, astronomers who scour the universe for exoplanets—planets orbiting other stars—think these two ice giants, though outliers in our solar system, may exemplify our galaxy's most common type of planet.

PLUTO AND PLANET X

In 2006, Pluto joined the fate of Ceres, Pallas, Juno, Vesta, and the many other demoted "planets" that preceded it—another edit in the continually rewritten story of our Sun's backyard.

Pluto is not a planet. It never was. But for 76 years it was officially listed as our ninth and most distant neighbor. And when the amateur astronomer Clyde Tombaugh discovered it in 1930, Pluto was heralded as the long-lost Planet X.

Recall the deluded Percival Lowell, who claimed to see Martian canals built by intelligent aliens. After his fantasies were dashed by observational evidence, he dedicated the rest of his life to the pursuit of Planet X—a planet that would account for the unexplained deviations in Neptune's orbit. At Lowell's observatory, 14 years after his death, Tombaugh discovered the long-predicted planet—the one that Lowell had reasoned must be at least half the size of Neptune to influence it so visibly. By the late 1970s, however, decades of observations gradually reduced the estimations of Pluto's total size and mass until finally, the poor thing was deemed far too small to exert any meaningful influence at all over Neptune's motion. And so, another hunt for a lost world continued unfulfilled.

Planet X would ultimately be killed, but in an entirely different way from planet Vulcan. The Voyager flybys provided updated and far more accurate measurements of the mass of the outer planets. In 1993, E. Myles Standish, Jr., realized that the long-accepted measurements of Neptune's orbit were awry. The combination of Voyager 2's refinements of the planet's mass and Standish's corrections of its motion fully accounted for Neptune's 164-year orbit. Planet X was not a planet at all, but rather poor calculations left unverified.

The year before the fall of Planet X, astronomers David Jewitt and Jane Luu found another object, about one-fifteenth the diameter of Pluto, orbiting beyond Neptune. Their discovery was followed by many more. Soon it became clear that Pluto resided in a large ring of debris, a new plot of real estate in our solar system. It was named the Kuiper belt, after Dutch astrophysicist Gerard Kuiper, who decades earlier had first hypothesized about icy bodies beyond Pluto.

In the years to follow, Pluto faced growing scrutiny as larger and larger objects were detected among the untold millions of chunks in orbit within the Kuiper belt. Finally, the time of reckoning arrived. In 2003 American astrophysicists Michael Brown, David Rabinowitz, and Chad Trujillo discovered Eris, an object more massive than Pluto (though a teensy bit smaller), with at least one moon of its own. There was little reason to doubt that additional, even larger, planetlike objects lurked beyond, simply awaiting our discovery. Either all these objects were planets, or none were. The seemingly obvious yet intractable question had to be answered: What, exactly, is a planet?

> Either all these objects were planets, or none were. The seemingly obvious yet intractable question had to be answered: What, exactly, is a planet?

Facing pressure to decide, the International Astronomical Union put the question to a vote in January 2006. Seven months later, by an overwhelming consensus, the members declared that Pluto, as well as any other Pluto-size objects yet to be discovered within the Kuiper belt, would henceforth be deemed dwarf planets.

Pluto (front) and its moon Charon's dynamic surfaces, captured for the first time by NASA's New Horizons spacecraft in 2015, are seen here in this high-resolution enhanced color view.

To achieve planethood in our solar system, one must check all three of the following boxes:

1. Are you a sphere?
2. Do you orbit the Sun?
3. Do you (and here is where Pluto failed) clear your orbit? In other words, is the mass of the debris within your orbital real estate less than your own total mass?

As for number 3, clearly the total mass of the Kuiper belt far outweighs that of Pluto.

While Pluto was receiving its final judgment, a first-of-its-kind mission toward the dwarf planet had already been en route for half a year. The New Horizons spacecraft—weighing in at about a thousand pounds and launched off Earth by a super-powerful Atlas V rocket at a record speed—zipped past the Moon after fewer than nine hours. (Apollo astronauts required more than three days to cross that same distance.) Along with a suite of imaging tools and experiments, Clyde Tombaugh's ashes rode the spacecraft toward his discovery and beyond.

Considering Pluto's distance from Earth, which averages more than three billion miles, getting there before the New Horizons scientists die of old age has required extra help from the planets. After a few gravity assists and a nine-and-a-half-year journey, New Horizons streamed past Pluto at 45,000 miles an hour in 2015. There, it found ice mountains the size of the Rockies and strong hints of a deep slushy, watery ocean—perhaps, just perhaps, capable of hosting life.

ON TO BEYOND

New Horizons later joined four other spacecraft en route to interstellar space: Pioneers 10 and 11 and Voyagers 1 and 2. The definition of interstellar space remains as elusive as the question of where Earth's atmosphere begins and ends. We know that the Oort cloud—a large, spherical shell of trillions of icy comets—lies well beyond where our farthest spacecraft have reached. At its current velocity, logging more than 900,000 miles a day, Voyager 1—the fastest human-made thing in the universe—will need another 300 years before it arrives at the outskirts of that shell, and another 30,000 years to traverse it. By then, the Voyagers, Pioneers, and New Horizons will be long dead, their metallic remains an ode to the ambitions of humanity.

What would the story of our solar system look like if we wrote it in ink on some imperishable paper as we went along, passed down from Aristotle to the astronomers of today? It would appear as an ever growing list of discoveries and breakthroughs, nearly illegible with revisions, edits, and deletions, the surrounding margins brimming with scribbled queries. In the unmapped reaches beyond those margins lie manifold stories yet to be told.

PART 3

INTO OUTER SPACE

"I do not know what I may appear to the world; but to myself I seem to have been only like a boy playing on the sea-shore, and diverting myself in now and then finding a smoother pebble or a prettier shell than ordinary, whilst the great ocean of truth lay all undiscovered before me."
—Isaac Newton, shortly before his death

We speak of space exploration, space travel, and spacecraft without pausing to ask the question: What's space? Wedged between galaxies, stars, planets, molecules, and atoms, space is not mere nothingness. In this section, as our cosmic journey continues, we escape the familiar—our solar system. We may plummet through darkness, but we certainly will not encounter nothingness. In fact, we will discover empty space is teeming with stuff—a roiling soup of quantum particles, even in the darkest and coldest corners of the universe.

Humanity's ideas about where and what space is began with the sky—blue and vast and mysterious as the seas. We called it the aerial ocean. At various times in history, descriptions and assumptions about the sea were bestowed on the sky and, later, on the entirety of space beyond.

We never let go of this comparison. Thirteenth-century Norse mythology relays the tale of the *Skidbladnir*, a magic ship that could be carried through the sky as easily as through the seas. Arthur Conan Doyle's 1913 short story "The Horror of the Heights" features giant flying jellyfish that terrorize aeronauts. In Disney's 1953 animated film *Peter Pan,* Captain Hook's pirate

The *Jolly Roger* pirate ship defies gravity to race
skyward in Disney's *Peter Pan* (1953).

PREVIOUS PAGES: In 2022, the Hubble Space Telescope captured this image
of a swirling cloud of gas and dust cocooning a young star in the
Taurus constellation more than 9,000 light-years away.

ship, the *Jolly Roger,* sails Wendy home from Neverland through the clouds. Whales inhabit the stratosphere and outer space in *Fantasia 2000.* The list could fill a book all its own.

And this cultural influence is felt in the real world. Today, after all, we fly to space in rocket *ships,* not rocket cars or rocket trains; even the International Space Station has a port side and a starboard side. In sci-fi films, rocket ships are often called boats, and the fictional dark unknowns they traverse are sometimes populated with creatures not unlike those that live in the sea, with large, lidless eyes, willowy bodies, sprouting tentacles, and green or blue skin.

Romantic comparisons between outer space and Earth's oceans are irresistible. Both realms are profoundly mysterious: dark and cold and vast beyond imagination. And yet, the two expanses can be complete opposites: a unique connection in itself. Examples abound. Deep-sea divers require special suits to combat extremes of atmospheric pressures, while astronauts require suits to combat the exact opposite scenario. Without the encasing bubble of Earth's atmosphere, both aquanaut and astronaut will meet a swift death.

Today, *Homo sapiens* have mastered navigation across both the briny deep and the sky above. But within the aerial ocean of outer space, we remain no more than a Moon-faring species; we've as much right to claim the term "spacefarers" as the "sailor" who dips a single toe into a shallow tidal pool.

The boundary between outer space and not outer space, as we know by now, is generally defined by the Kármán line—where aeronautics becomes astronautics. On Earth, it's about 100 kilometers (62 miles) above sea level. But what about on Jupiter? Or Pluto? The boundary between anything in the universe and some other thing can be defined only by conventions.

When you hold the tip of your index finger to the tip of your thumb, for example, you can see a defined boundary between

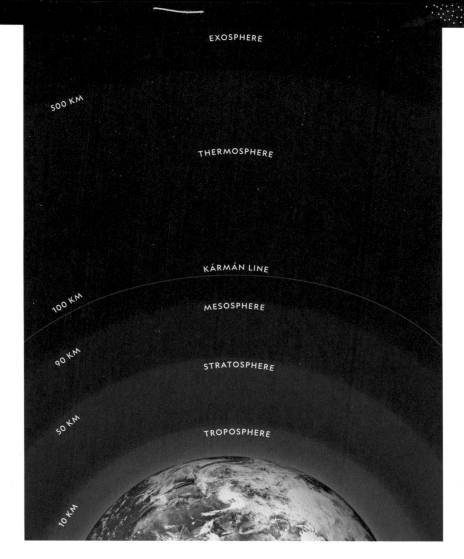

EXOSPHERE

500 KM

THERMOSPHERE

KÁRMÁN LINE

100 KM

MESOSPHERE

90 KM

STRATOSPHERE

50 KM

TROPOSPHERE

10 KM

A schematic representation of the five principal layers in Earth's atmosphere, from the troposphere and stratosphere to beyond the Kármán line, the delineation between our atmosphere and space

the two phalanges. If you press the two together, you can now also feel the boundary. But if you were to observe the separation with a powerful microscope, gradually increasing the magnification, the boundary you'd thought firm and defined would become increasingly vague. The ridges and divots of your skin would overlap in some places and not others. The bacteria of one finger might migrate to the other finger, along with tiny molecules of debris or lotion, each instance changing the total mass of the separate fingertips. If you could look even closer, you might witness the electrons between the atoms of both fingers repelling one another, leaving the tiniest gap no matter how hard you squidged them together. How might you define the boundary now? Eventually, you'd just accept that the boundary between your thumb pad and forefinger is simply where you can no longer fundamentally define which is which.

And keep in mind this little exercise involved two solid objects. What about assessing the boundaries of nonsolid things, such as the vacuum of space?

INTO THIN AIR

"You're not the same as you were before," says the Mad Hatter in the Disney film *Alice in Wonderland*. "You were much more . . . muchier . . . you've lost your muchness." In our attempts to define outer space, we sound like the Mad Hatter admonishing Alice for losing her muchness. We tend to define space not by what it is and has, but by what it is not and doesn't have. It has fewer particles than Earth's atmosphere. It has less pressure. It is an absence of heat, of light. It's enticing to believe that by just venturing farther and farther from Earth, we will eventually arrive at a place of total emptiness, devoid of all things. We often think of space this way—as a dark abyss, as nothingness. But, no. Space is not nothing, and it's far from empty, as we will discover in this section.

Often our declarations of what exists or does not exist derive from the five fallible senses with which we perceive the environment. Our imperfect sensory interface with the physical world limits our attempts to understand the universe every day. Everybody knows that air is thin; why else would we be impressed when a magician pulls a rabbit "out of thin air"? Yet a thimble-size container (one cubic centimeter) of pure, invisible, sea-level air contains more molecules (27 quintillion) than all the grains of sand on an average beach.

Is this a lot or a little? A rabbit that pops in and out of an otherwise empty magician's hat may be interested to know that the average density of matter in the universe is much, much less than the air in that hat. But how much less?

Let's upgrade our thimble-size container to a hot tub–size (35 cubic feet) birthday present with your name on it. If you unwrapped your gift only to find no objects inside, you would justifiably feel the giver was a bad friend for gifting you a big box full of nothing. But what your friend gave you was 2.7×10^{25} molecules of air. That's 27 septillion molecules (27,000,000,000,000,000,000,000,000). You can't fairly claim they gave you nothing when they gave you 27 septillion somethings. (Pocket this little fact for later, in case you ever forget a special someone's birthday and find yourself in need of a quick present.)

The rarest, most mythical, most sought-after, and pondered-over gift in the universe would in fact be nothing— true nothing.

So, let's say your friend now wants to make up for their lousy gift of 27 septillion molecules and is determined to give you the gift of true nothing. Might they have better luck finding it in outer space? If they could somehow obtain a one-cubic-meter sample of the space between planets, box it up, and bring it back to Earth, you might believe the box to be truly empty. Again, no. Inside that box (assuming it to be a magic box, impervious to the crushing differences in atmospheric pressure) would be perhaps five million somethings, mostly hydrogen and charged particles of solar wind.

Certainly, the farther away from the solar center anybody journeys in search of nothing, the fewer and fewer particles they'll encounter, approaching nearer and nearer to zero. Beyond the solar system and out there between stars (interstellar space), they'll find about a million particles per cubic meter. Beyond the Milky Way and between galaxies, in intergalactic space, a few hydrogen particles will still be floating around within each cubic meter.

Okay, but surely in the space between two meandering intergalactic particles one might find true emptiness, true nothing, mightn't one? Wrong again. Out here, the weird laws of quantum physics reign—with peculiar particles all their own.

Meet the virtual particle. No, it's not the kind of matter you'd find in an alternate universe. Virtual particles are the real magician's rabbits of the universe, popping into and out of existence,

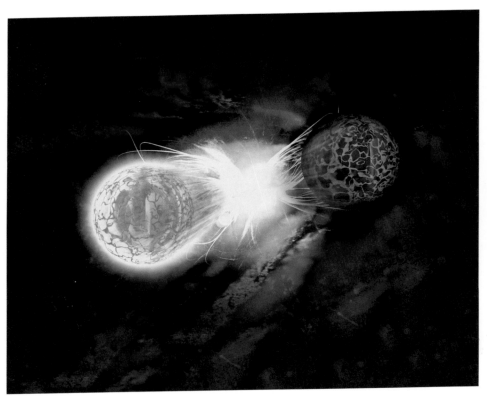

This computer-generated artwork depicts a particle and an antiparticle being created in a vacuum in a process known as quantum fluctuation.

out of thin air—indeed, the thinnest of air, which we might instead call "thin space." Heisenberg's uncertainty principle of quantum physics forces us to accept that the space between ordinary particles seethes with virtual particles that pop in and out of existence, undetected. As a result, we cannot call that space empty.

Even between particles, therefore, we find particles. And permeating all of space and time, between the particles and virtual particles, is the inescapable influence of gravity and multitudinous energy fields. True nothingness, a place of no things, does not exist—at least not in this universe.

The notion that outer space is a realm of nothingness is a relatively recent misconception. Aristotle famously declared in the fourth century B.C.E. that nature abhors a vacuum. On philosophical grounds, he reasoned that the definition of "nothing" meant "not being." Nevertheless, nothingness exists as a concept that can be discussed. *Not* a thing, in other words, was still a thing to be.

On a physical level, Aristotle reasoned, the surrounding air would immediately rush in to fill any void. Furthermore, if such a void could exist, it would mean that matter could travel infinitely fast within it. Yet infinitude could not exist, because the heavens were the biggest and vastest entity: immense, but finite. There simply wasn't room for something infinite.

> The notion that outer space is a realm of nothingness is a relatively recent misconception . . . Nevertheless, nothingness exists as a concept that can be discussed.

Aristotle also believed that the heavens and the wandering lights that revolved around us were made of a fifth substance, or quintessence, distinct from the four terrestrial elements:

continued on page 176

VACUUMS AND VOIDS

The best vacuum we know of in the solar system resides on Earth—or rather, inside Earth. Three hundred feet below the village of Meyrin, Switzerland, sits the Large Hadron Collider, which holds fewer particles than the space between planets. Inside the 16-mile evacuated circular chamber—operated by CERN, the European Organization for Nuclear Research—particles shoot through pipes and collide at nearly the speed of light, scattering shrapnel of smaller particles in the aftermath.

To ensure these particles hit only one another and not any stray atmospheric molecules or vagrant atoms, the inner chamber must contain as close to nothing as possible. To create such plentiful nothing, the collider's specially coated pipes absorb any undesirable molecules that may have parked on their surface, and the vacuum chambers bake at almost 500°F. Then, two weeks are spent pumping gas into the vacuum system, supercooling it and drastically reducing its atmospheric pressure to eliminate as many other stray molecules as possible. The collider also hosts the coldest place in the solar system, making CERN humanity's greatest achievement in the fields of cryogenics, vacuum technology, and particle physics—as well as the closest we get on Earth to nothingness.

Beyond the solar system, space becomes more and more rarefied. But does a specific region of the universe warrant the description "nothing-est"? Or, as the Mad Hatter might say, the least "muchy"?

That honor likely goes to the Boötes void, a region of space 700 million light-years away. At 300 million light-years across, it's often called the Great Nothing. It hosts 60 known galaxies, but anywhere else in the universe, a region of that volume would host at least a couple thousand. By comparison, then, it's quite empty. Tens of millions of light-years separate each of the Boötes's galaxies, whereas the Milky

Way's nearest neighbor, the Andromeda galaxy, is only two and a half million light-years away. Thus, it's likely that the space between the Boötes void galaxies contains the lowest particle density in the observable universe. If you found yourself floating there, you'd be as lost in the sea of space as anybody could be, adrift amid utter darkness with not a single beacon star to guide you.

The Large Hadron Collider, CERN's flagship accelerator, during a period of upgrades known as the second long shutdown in 2019–2020

An engraving of the Greek philosopher Aristotle (384–322 B.C.E.) from *The Trek of Physics* by Thomas Bricot (1496)

continued from page 173

earth, air, fire, and water. This quintessence, perfect and heavenly, which he proposed be called aether, made up the celestial realm and remained a core feature in the geocentric, Ptolemaic view of the universe.

During the scientific revolution, although a Sun-centered universe came to replace the Earth-centered one, the concept of a heavenly aether was simply redefined as the stuff that fills all space between planets and stars. Its new name was the luminiferous aether. This refinement conveniently solved the question of how light waves could possibly travel through "empty" space without a medium to transmit them. Now there was a medium—the luminiferous aether—which permeated the nothingness of space, just as molecules permeate the nothingness of air. This put the nature of light, in the universe and here on Earth, front and center in the thoughts of physicists.

LIGHT: WAVE OR PARTICLE?

Too often, humans take an either-or approach to information and knowledge. We require that something (or someone) be one thing or another, firmly categorized and jammed into a bucket of prescribed features. These urges tend to thwart scientific, cultural, and social enlightenment. But, in fact, a common pattern emerges throughout the history of science: If observations and measurements lead us to think that two different answers to the same question may be true—even if they are contradictory—then the right answer may simply be some combination of the two results, not one exclusive of the other.

The nature of light offers a perfect historical example of such a pattern. Today, we understand that a photon—an individual packet of light—behaves as both a wave and a particle (a wavicle, if you will, although that word has never caught on).

At first, physicists thought it was either one or the other. Around the turn of the 18th century, Christiaan Huygens and Isaac Newton championed two opposing claims about the fundamental properties of light. Newton proclaimed light to be a beam of particles—corpuscles, he called them—while Huygens insisted that light traveled as a wave, not fundamentally different from how sound propagates through a medium.

Both and neither were correct. Light as waves was an easy sell. But we'd have to wait two centuries before Albert Einstein experimentally demonstrated and persuasively described light as particles, earning him a Nobel Prize.

LIGHT, A WAVE

If both sound and light travel as waves, then maybe they're more similar than different. Christiaan Huygens's *Treatise on Light*, published in 1690, proposed that light behaves like waves.

Learned people knew that sound cannot travel through air-less voids. Irish philosopher Robert Boyle, inspired by Evange-lista Torricelli's new barometer (described on pages 31–33), had conducted dozens of experiments with vacuum chambers. Among his discoveries was the odd behavior (or lack thereof) of sound in an airless void. But light, he found, traverses the vacuum without hindrance. So, what medium propagates light waves in the absence of air?

To answer this question, Huygens called upon the aether, the hypothetical invisible substance permeating the air around us and the space above. "One sees here," Huygens wrote in his 1690 *Treatise on Light*, "not only that our air, which does not penetrate through glass, is the matter by which Sound spreads; but also that it is not the same air but another kind of matter in which Light spreads." Furthermore, he contended, "It is not beyond the limits of probability that the particles of the ether have been made equal for a purpose so important as that of light, at least in that vast space which is beyond the region of atmosphere and which seems to serve only to transmit the light of the Sun and Stars."

Huygens had backed himself into a corner, forcing the hypothesis that the aether must be as fundamental to the propagation of light as air is to the propagation of sound. But by invoking an aether in which light waves propagate, Huygens solves the conundrum of how light travels in a vacuum and how light from the Sun and distant stars reaches earthly eyes.

Huygens's wave theory fell short of satisfactorily explaining two fundamental behaviors of light: that it cannot bend around walls the way sound waves can, and that it travels in a straight line. You can even try it yourself. Cut a small pinhole through the exact same spot on three rigid pieces of cardboard and prop them up in a straight line several inches apart so that you can peer down the sight line of all three holes as though you're looking down a gun barrel. Place a candle or small lightbulb directly in front of the three cardboard pieces. If you shift any of the three, you'll disrupt your sight line, and the light will no longer reach your eye. There's your proof: Light travels in a straight line.

LIGHT, A PARTICLE

Before astrophysics, there was astrology. Before chemistry, there was alchemy. In our day, alchemists are thought of as gold-greedy magicians, but at the core of their practice were two ideas: that two ingredients can combine to form a third substance, and that there is a singular, smallest particle that makes something what it is.

Many people forget that Isaac Newton, in addition to his work in physics and philosophy, was also an avid alchemist. He wrote an estimated one million words on the subject—as least as much as he wrote about mathematics. Perhaps influenced by his alchemical endeavors, he believed that light, like all other matter, could be separated into smaller constituent particles. Particles, or corpuscles—unlike waves—do not require aether, or indeed any medium at all, to propagate. As Newton knew, you cannot hear a bell ringing inside a vacuumed-out glass—but you can still see the bell. Light has no trouble crossing the vacuum.

Newton's particle theory of light, as laid out in his 1704 treatise, *Opticks,* explained why light can travel in a straight line, and why it reflects off a mirror at the same angle it arrived. Because more tangible objects, such as a bouncing ball, behave the same way when tossed at a hard surface, the theory seemed compelling and was widely accepted.

But that wasn't the full story. Why, when sunshine beams through a prism, did a rainbow of red, orange, yellow, green, blue, indigo, and violet always appear, and always in that exact order? Newton argued that red light contained larger corpuscles than orange, and orange larger than yellow, and so on. The variations in refrangibility produced the colors. The corpuscular theory of light explained the mysteries of the rainbow as well as the reason light travels in a straight line and cannot bend around walls.

For nearly a century, Huygens's wave theory was largely dismissed, overshadowed by young Newton's sway in the scientific community. Eventually, though, the British polymath Thomas Young elaborated on Huygens's theory to demonstrate that waves could allow light to propagate in a straight line, again invoking the aether as fundamental to the explanation.

For the next hundred years, the aether was as foundational to our understanding of our universe as the geocentric model had once been. British mathematician Lord Kelvin (for whom the absolute temperature scale is named, and who formulated the first and second laws of thermodynamics) dedicated lavish mathematical efforts toward accounting for the necessary properties of the aether. All told, during the era of modern science, the aether was perhaps the greatest assumption to live long and prosper without ever having had a scrap of observational evidence to support it.

THE DEATH OF THE AETHER

"Whatever difficulties we may have in forming a consistent idea of the constitution of the aether, there can be no doubt that the interplanetary and interstellar spaces are not empty, but are occupied by a material substance or body, which is certainly the

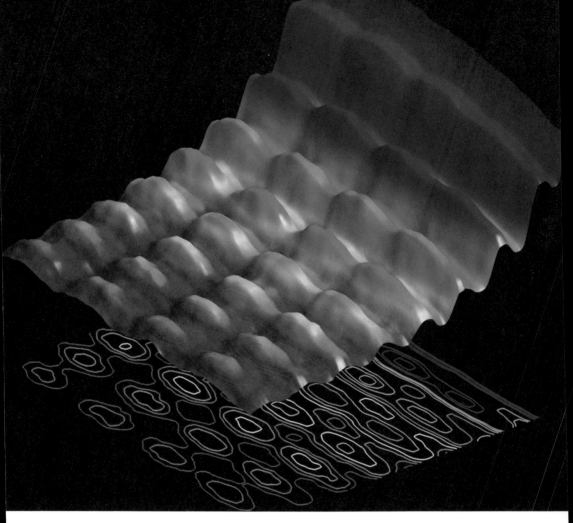

The first photograph ever taken of light visible simultaneously as both spatial interference and energy quantization—as both a particle and a wave

largest, and probably the most uniform body of which we have any knowledge," wrote James Clerk Maxwell in 1878.

Maxwell was right. Space is not empty—but not in the way he imagined. By the late 19th century, scientists began to wonder: If there is an aether, then we should be able to measure its effects on the speed of light as Earth moves through it. The speed of

continued on page 184

WHALE SONGS AND SUNSETS

hotons streaming from our Sun cross the quasi-vacuum of space at—you guessed it—the speed of light. Once these teensy packets of light encounter Earth's atmosphere, they slow down. Whenever light slows down, its path bends—it refracts. Observe a straw in a clear glass of water. It kinks at the boundary between the air and the water's surface because light travels through water more slowly than through the air above.

When observed through our atmosphere, sunlight behaves like the straw. We do not see the Sun where it is, but where the refraction of light from empty space into Earth's atmosphere allows us to see it. Near sunset, with the Sun low in the sky, sunlight passes through even more air—multiple atmospheric layers more. The resulting refraction is so profound that by the time you see the lower edge of the setting sun kiss the edge of the horizon, the Sun has already set. Aim a missile at the setting Sun, and you will miss your target by a long shot. The same holds for the sunrise. We see the Sun where it is yet to arrive. Each day, the refraction from both sunrise and sunset offers us a few more minutes of daylight than we've otherwise earned.

The behavior of light offers no insights regarding the behavior of sound through media. They are fundamentally different, and we were distracted for centuries thinking otherwise. Not only does sound not travel through a vacuum; it travels the most slowly through gas, a bit faster through liquid, and fastest through metals.

Back in grade school, perhaps you tied a long string between a pair of tin cans or paper drinking cups to communicate secret messages to your buddy a few desks away. In that game, you were tapping into the physics of sound waves. The bottom of the cup absorbed your sounds, transmitted them across the string, and deposited them to the bottom of your friend's cup, re-creating your voice. Whales evolved their own version of string-can communication.

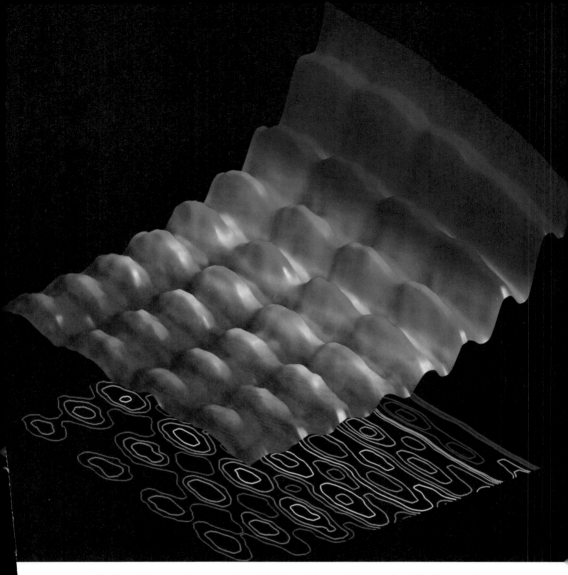

The first photograph ever taken of light visible simultaneously as both spatial interference and energy quantization—as both a particle and a wave

largest, and probably the most uniform body of which we have any knowledge," wrote James Clerk Maxwell in 1878.

Maxwell was right. Space is not empty—but not in the way he imagined. By the late 19th century, scientists began to wonder: If there is an aether, then we should be able to measure its effects on the speed of light as Earth moves through it. The speed of

continued on page 184

WHALE SONGS AND SUNSETS

Photons streaming from our Sun cross the quasi-vacuum of space at—you guessed it—the speed of light. Once these teensy packets of light encounter Earth's atmosphere, they slow down. Whenever light slows down, its path bends—it refracts. Observe a straw in a clear glass of water. It kinks at the boundary between the air and the water's surface because light travels through water more slowly than through the air above.

When observed through our atmosphere, sunlight behaves like the straw. We do not see the Sun where it is, but where the refraction of light from empty space into Earth's atmosphere allows us to see it. Near sunset, with the Sun low in the sky, sunlight passes through even more air—multiple atmospheric layers more. The resulting refraction is so profound that by the time you see the lower edge of the setting sun kiss the edge of the horizon, the Sun has already set. Aim a missile at the setting Sun, and you will miss your target by a long shot. The same holds for the sunrise. We see the Sun where it is yet to arrive. Each day, the refraction from both sunrise and sunset offers us a few more minutes of daylight than we've otherwise earned.

The behavior of light offers no insights regarding the behavior of sound through media. They are fundamentally different, and we were distracted for centuries thinking otherwise. Not only does sound not travel through a vacuum; it travels the most slowly through gas, a bit faster through liquid, and fastest through metals.

Back in grade school, perhaps you tied a long string between a pair of tin cans or paper drinking cups to communicate secret messages to your buddy a few desks away. In that game, you were tapping into the physics of sound waves. The bottom of the cup absorbed your sounds, transmitted them across the string, and deposited them to the bottom of your friend's cup, re-creating your voice. Whales evolved their own version of string-can communication.

Sound waves travel about four times faster through the ocean than through the air above. Under the right conditions, whales can sing to one another separated by thousands of miles. Pressure and temperature differences within the ocean cause boundary layers that channel the propagating sounds. This layer traps sound waves, bending and reverberating them across thousands of miles underwater. Think of it as an underwater tunnel for whale songs, allowing their communication frequencies to remain clear and strong over long distances.

Dolphins and other underwater mammals also exploit the quick propagation of sound through water for echolocation. High-pitched clicks bounce off objects and back to their ears, imparting information about the position, size, and movement of whatever or whoever is in the vicinity. By using sound through liquid to sense the environment, underwater echolocators can perceive their surroundings better than any animal that relies upon light passing through air.

A humpback whale off the coast of Tonga

U.S. physicist Albert Abraham Michelson (1852–1931) with one of his interferometers at the U.S. Naval Academy in Annapolis, Maryland

continued from page 181

light should be different, they thought, when light moves in the same direction as Earth moves in its orbit (as opposed to moving at right angles or in the opposite direction as orbiting Earth).

Imagine tossing a golf ball forward from an open window of a speeding car. At that instant, a person standing by the side of the road would measure the golf ball's speed through the air as the speed of the car plus the speed with which the person threw the ball. Now repeat the experiment, but this time throw the golf ball backward. The curbside measurement of the golf ball's speed through the air is now the speed of the car minus the speed at which you tossed the ball.

Prussian-born American physicist Albert Michelson joined the U.S. Naval Academy at the age of 17, imbibing nautical knowledge of wind and sea as he began his scientific studies. That connection may have influenced his research into wind of the

celestial kind—ethereal wind. Michelson invented a specialized instrument called the interferometer, capable of measuring the velocity of light beams with exquisite precision.

In 1887, Michelson and the American chemist Edward Morley conducted several experiments designed to measure the precise velocity difference of light through the cosmic aether. Alas, the fundamental premise of their experiments was flawed. The velocity of light never changed, not once, no matter which direction the light moved relative to Earth. No medium changed the speed, which forced scientists to conclude that the mythical aether did not exist—or that if it did, it held no influence over light.

Their efforts are still considered the most famous botched experiment in the history of science. Through this magnificent failure, Albert Einstein found the necessary footing to postulate his 1905 special theory of relativity, which, together with his 1915 general theory of relativity, put the final nail in the coffin of the aether and ushered in a new era of thinking about the cosmos. One might wonder: If not for the Michelson-Morley experiment, might Einstein have lived and died without having advanced his theories of relativity? And if he did not, would anyone else have ever done so?

Special relativity, which describes the movement of mass and energy through spacetime, was built upon the equations of Scottish mathematician James Clerk Maxwell. Maxwell's equations showed that light is a disturbance of the conjoined electric and magnetic fields, propagated by the aether. Yet this newly understood electromagnetic phenomenon called light required no medium at all to propagate.

GRAVITY AND LAGRANGE POINTS

To ask, "What is space?" is to also ask about gravity. Gravity is inescapable. The term "zero g," which ostensibly means "no

gravity whatsoever," is a misnomer and fosters a persistent misunderstanding of how this force works. Weightlessness— the experience we dub "zero g"—is only possible *because* of gravity: the outcome of one object freely falling toward another, literally succumbing to its gravitational pull.

All objects exert a force of gravity that extends to infinity, forming the very shape of the spacetime continuum. Einstein's general theory of relativity, simply put, tells us that matter and energy distort space in their vicinity, behave the same way, and are indistinguishable from each other in their effects. Einstein most famously declared that $E = mc^2$: The energy of an object equals the mass multiplied by the speed of light, squared. The larger the chunk of stuff, the greater the gravity and the greater the bend in the fabric of space and time. American physicist John Archibald Wheeler summarized this concept perfectly in his book *Geons, Black Holes and Quantum Foam:* "Spacetime tells matter how to move; matter tells spacetime how to curve."

Envision two heavily weighted balls resting atop a suspended rubber sheet. The heavier the ball, the deeper the valley formed around it. There exists a precise point on that sheet between those two bodies where you could place a third, lighter ball that would, albeit precariously, remain at rest and in balance between the two perfectly canceled out "gravitational pulls." This third ball now falls toward neither of the other two balls; instead, it stays at the peak between the two valleys created by the heavier balls. Five such balance points, where all forces are equalized, exist for any pair of orbiting objects.

Astrophysicists call these special places in space Lagrange points, after the Italian mathematician Joseph-Louis Lagrange, who first derived their existence in 1772. The Lagrange points encapsulate the interaction of gravity and matter in the fabric of spacetime. We exploit this fact of physics when we park space

telescopes and satellites beyond Earth. In the environment of any two orbiting objects lie five different points of Lagrange.

The first point of Lagrange, or L1, lies between the two bodies—just like that third ball on our rubber sheet. An object parked at L1 in the Sun-Earth system maintains a continuous view of either the Sun or of a fully lit Earth while orbiting the Sun at L1, in lockstep with Earth. From this vantage point, the Solar and Heliospheric Observatory (SOHO) satellite returns a continuing stream of data about the Sun's composition and behavior, providing an important defense against potentially disastrous solar flares and coronal mass ejections. Also in orbit at the Sun-Earth L1 is the National Oceanic and Atmospheric Administration's Deep Space Climate Observatory (DSCOVR), which continuously monitors both space weather and Earth's changing climate.

Our second Sun-Earth point of Lagrange, L2, lies on the opposite side of Earth from L1—not between Earth and the Sun, but on the far side of Earth. Why doesn't such an object simply fall immediately toward Earth, doubly influenced by the pull of our planet plus our star? Because these points are located within a rotating system, not between stationary balls on a rubber sheet. Recall centrifugal force, the urge to fly outward from the spinning merry-go-round (see page 49). Any object in orbit around another object experiences this same urge to fly outward from the rotating system. Lagrange points balance this outward force with the gravitational pull exerted by both objects. At L2, the outward centrifugal force on an object is

Astrophysicists call these special places in space Lagrange points, after the Italian mathematician Joseph-Louis Lagrange . . . The Lagrange points encapsulate the interaction of gravity and matter in the fabric of spacetime.

strong, but the Sun's and Earth's inward gravitational pull tether the object.

At a million miles away, the Sun-Earth L2 is a terrible place for satellites to observe Earth, which is permanently in the Sun's glare. However, looking the other way, away from both Sun and Earth and into the depths of unilluminated space, it's the perfect location from which to observe the rest of the universe. The most expensive telescope ever constructed, the $10 billion James Webb Space Telescope (JWST), resides at the Sun-Earth L2, where it can observe and relay back information about the most distant, most ancient parts of the observable cosmos.

L3, the third Lagrange point of the Sun-Earth system, lies way over on the opposite side of the Sun from L1 and L2. As far as we can tell, this point serves no useful purpose beyond science fiction plot devices. Remember the planet Vulcan that never was? Some sci-fi writers speculate that an undiscovered Planet X resides at L3, ever hidden behind the Sun in perfect synchrony with Earth's rotation.

Here's the problem with a hypothetical Planet X or in fact with any object located at the first three points of Lagrange: Their stability cannot last forever. Their precarious balance means that the slightest shift—say, the subtle tugging of seven other planets in the solar system—could easily knock such an object out of position, just like what happens to a ball balanced on the top of a hill when a gust of air hits it. To stay in their parking zones, the SOHO and JWST satellites must make continuous minor adjustments, using onboard fuel carried for just this purpose. Within a couple generations of human observation, any "hidden" Planet X would drift out of orbit and into view.

This graphic illustrates the Lagrange points for a star and an orbiting planet. There are five such points where the gravitational and orbital forces of the two bodies balance out.

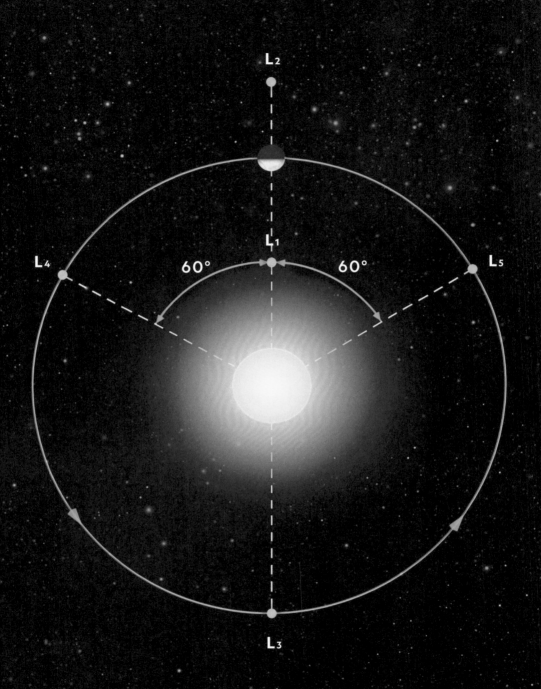

The most stable parking spots are found at L4, which lies ahead of any orbiting object, and L5, which trails the object at an equal distance. If L1, L2, and L3 are like balls atop a hill, then L4 and L5 are like balls resting in the valley between two hills. Unlike a nudge to a ball at the top of a hill, a nudge to a ball at the bottom merely shifts it a bit before it spontaneously returns to the same position as before.

L4 and L5 are thus supremely desirable as a parking spot for observational craft. They require little to no fuel to remain in position. They're convenient depositories not only for human-made hardware, but also for asteroids and other loitering space debris.

In the Sun-Jupiter system, the L4 and L5 Lagrange points have collected thousands of asteroids—at least one of which measures more than a hundred miles in diameter. These trapped rocks are called the Trojan asteroids, after the ancient war depicted in Homer's *Iliad*. The larger asteroids found at L4 are traditionally named for ancient Greek heroes, such as 1143 Odysseus, and at L5, for the heroes of Troy, such as 1208 Troilus.

SHOCKING TRUTHS

The luminiferous aether hypothesis, killed swiftly by the Michelson-Morley experiment and by Einstein's relativity theory, was an attempt to explain what 17th-century scientists knew and 21st-century Hollywood directors refuse to accept: In space, nobody can hear you scream (though they can watch you die). The demise of this elusive quintessence should have also informed said Hollywood directors about the behavior of explosives in a vacuum.

Down here on Earth, if you want to blow up something with a conventional bomb, you need a medium and a shock wave generated by exploding the bomb within the medium. Bombs

don't just puff a gust of air, like the wolf in "The Three Little Pigs." The intense heat generated at the instant of detonation creates a ferocious pocket of expanding air that moves faster than the local speed of sound: the precise recipe for a shock wave. It isn't the heat of an explosion or flying shrapnel that kills the most people and razes entire cities to the ground. It's the shock wave—the catastrophic imbalance of pressure between the sides of nearby structures (and people) facing the explosion and the sides facing away. Such are the forces that blow things apart into unrecognizable fragments.

In the vacuum of space, earthly bombs are useless, at least insofar as shock waves go; there aren't enough molecules in the medium to transfer the energy from one to the next. But as we know, the cosmic vacuum is never truly empty—and even in such a low-density environment, an explosion with sufficient matter and energy can wreak havoc in the universe. Huge flares rear up out of the Sun and launch billions of tons of plasma into the solar system at a million miles an hour; gigantic rings of multimillion-degree gas race outward from exploding stars; colossal gas clouds and entire galaxies collide, plunging into one another and creating bursts of freshly formed stars.

Shock waves begin from a simple fact of nature: Molecules in a gas are always on the move. Depending on their shape, not only do they stretch and shrink and twist and turn and vibrate; they also move, body and soul, from one place to another. In a split second, a meandering molecule bounces off neighboring molecules, transferring energy and momentum from one to the next, and so on across the gas.

Intruders, too, can shake up a batch of molecules that are otherwise jiggling away in blissful isolation. When a plane flies through the air, the solidly attached molecules of its nose cone tear through the gaseous molecules ahead of it, creating a huge ripple of pressure. The ripple passes from the layer of molecules

closest to the plane, to the layer just ahead, and then to the layer just ahead of that, all at the speed of sound. Same thing happens ahead of the airplane's tail. As each layer feels the ripple, it bumps into the next layer, announcing the news of the oncoming plane. Meanwhile, the plane cruises through the air without incident.

But what happens if the plane flies faster than the time it takes for each layer to bump into the layer in front of it—faster, in other words, than the speed of sound? At the frigid air temperatures of cruising altitude, the speed of sound is about 660 miles an hour. If the plane is suitably designed and powered, it just busts through the medium's hapless molecules. All the pressure waves, including the ones created by the noise of the plane's engines, now pile atop one another, greatly amplifying the resulting sound.

Meet the sonic boom.

A sonic boom is the audio track of a shock wave. Anyone who happens to be nearby will hear it loud and clear. Amplify a plane's shock wave by a factor of 10, 100, 1,000, and you're on your way to simulating the conditions of some common happenings in outer space.

Every syllable you utter sends its own sound wave—its own wave of pressure—rippling through the air. When you stand in one spot and talk nonstop, each wave you generate forms a sphere that is centered on your mouth and expands at the speed of sound.

But let's say you're both chatty and speedy. If, for instance, you start at the base of the Washington Monument and talk while you walk north toward the White House, each new sound

> All the pressure waves, including the ones created by the noise of the plane's engines, now pile atop one another, greatly amplifying the resulting sound. Meet the sonic boom.

Two Northrup T-38 Talon supersonic jets break the sound barrier, producing shock waves heard on the ground as a sonic boom.

you make comes from a new expanding sphere that rides closer than normal to the leading edge of the preceding sphere and farther than normal from its trailing edge. Of course, the faster you walk, the closer together are the leading edges of successive sound waves.

You experience this phenomenon, called the Doppler effect, every time a siren, car, or train whirs past. The pitch of the siren, horn, or train whistle sounds increasingly higher as it approaches and then increasingly lower as it recedes. Austrian physicist Christian Doppler described the effect in 1842, when railroad systems slowly began snaking their way across the countryside.

In 1845, a Dutch meteorologist, C. H. D. Buys Ballot, conducted a simple yet brilliant experiment to demonstrate the Doppler effect to anyone in doubt. He positioned one band of trumpeters on a train platform and another band of trumpeters aboard a train scheduled to pass by. Both bands were instructed

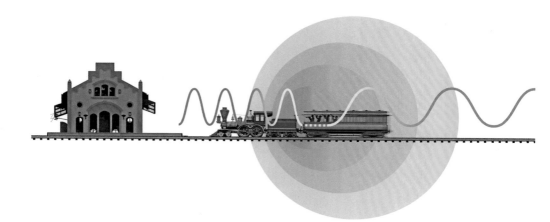

In 1845, Dutch meteorologist Christophorus Buys Ballot tested the Doppler effect by inviting musicians to play aboard a train. When heard from the station platform, the music was higher in pitch as the train approached (blue) and lower as it receded (red).

to play the same note at the same time. For listening onlookers, the steady, unchanging note of the platform band sounded drastically different than the higher-pitched then lower-pitched note of the band on the passing train.

Let's return to your chatty, speedy jaunt toward the White House. Suppose you walk so fast that the sound of your current syllable catches up with the sound of your previous syllable. If you keep walking and talking at that speed—the speed of sound—all your syllables will pile up together as you lay one track after another on the same leading edge. That would be your own personal shock wave.

When physicists refer to the speed of an object within a medium, they almost always invoke Mach numbers. This unit is named for the 19th-century Austrian physicist and philosopher Ernst Mach. By definition, an object that moves at the speed of sound moves at Mach 1. But don't ask, "How fast is that?" unless you're prepared to answer three questions: What is the temperature of the medium? What kinds of molecules make up

the medium? How compressible are those molecules? Those questions arise because, unlike the speed of light in a vacuum, which is the same anywhere in the cosmos, the speed that corresponds to Mach 1 is strictly local.

Nowadays, encounters with Mach 1 are not rare. The snap of a damp towel against your friend's butt at the gym is a mini sonic boom. So is the rapid inflation of your car's air bag. Want bigger booms? Try Mach 2 (the now retired Concorde commercial jetliner) or Mach 3 (the SR-71 Blackbird, a retired U.S. Air Force spy plane). And by the way, no matter the medium or the speed of sound within it, reaching similar Mach numbers creates similar physical phenomena.

Early in *Top Gun: Maverick* (2022), Tom Cruise's character ejects as a test pilot from a jet flying at Mach 10.5—approximately 7,000 miles per hour. In the next scene, he's calmly walking back to base. The hypersonic shock waves at that speed would have flattened him like a bug on a windshield. Just saying.

Ever experience a dish-rattling sonic boom? Most likely it came from a small, high-flying military aircraft. But if that plane had been large or flew supersonically at a low altitude, the boom wouldn't have been so innocent. Flown low enough, even an ordinary fighter jet can lay down a carpet of sonic booms that not only ruptures eardrums but also breaks windows and causes nosebleeds. As it reentered Earth's atmosphere at Mach 25, the returning space shuttle orbiter made two ferocious booms, one from the nose and one from the tail. Fortunately, the orbiter slowed to subsonic speeds before descending low enough for its booms to rattle anyone's brains.

Twentieth-century technology bequeathed its own versions of extreme shock waves here on Earth. Some were unleashed by the detonation of atomic bombs nicknamed "Little Boy" and "Fat Man" over Hiroshima and Nagasaki in August 1945. Others were let loose by the hydrogen bombs "Mike" and "Bravo"—hundreds

of times more powerful than their atomic predecessors—on the Pacific atolls of Enewetak in 1952 and Bikini in 1954. Dozens of other tests took place there between 1946 and 1958.

Unlike a conventional bomb, an atomic bomb doesn't need a medium to make it lethal: The high-energy light of the explosion itself passes straight through the transparent air. If you're a slab of concrete with a high melting point, you can easily survive this phase. But if you're an organic life-form and happen to be positioned close to ground zero, every molecule of your body burns, turning you to dust and vapor. Any surviving structures in the vicinity are then leveled by the stupendous shock wave moving through the medium.

One hesitates to contemplate the shock waves that might be wrought by 21st-century technology. As humanity enters a new era of spacefaring ambitions, so too will it enter a new era of war fighting. In December 2019, the U.S. government officially founded a new branch of the military, the Space Force, to entertain just such possibilities. This branch, as well as the redoubled military space efforts under way in many other countries, was conceived chiefly to think about what weapons might be used in space and how to defend against them.

The inextricable history of the (first) Cold War and the 1960s Moon shot, of World War II and the V-2 rocket, of World War I and the aeroplane all serve as stark reminders that aeronautical and astronautical technologies often advance in the service of war.

SHOCK WAVES BEYOND THE SOLAR SYSTEM

Out there beyond the solar system and between the stars, atoms and molecules are typically few and far between. Yet we know that the shock wave from a nearby exploding star created

our solar nebula, which then condensed and turned into our solar system.

How could this be? Nebulae are gas clouds. They contain the necessary molecules, and thus the medium, to generate shock waves—and with those waves, enough force to create a star. You don't have to look hard to find gas clouds actively engaged in the birth, life, and death of stars. The most extreme of these phases, along with the most spectacular shock waves, is stellar death.

Take a star with at least eight times the Sun's mass. Any star that massive is born fast, shines bright, dies young, and leaves a beautiful corpse. It spends its entire life in the fast lane. Eventually, though, its fuel runs low, and the fusion furnace at its core, which has kept the star from collapsing under its own weight, starts to shut down. At death, with no fuel left to fuse, the star swiftly implodes. The heat created by the precipitous collapse is so great that the entire wreck detonates in a titanic, multimillion-degree explosion that sends the star's outer layers bulldozing at hypersonic speeds into every gas cloud in the neighborhood. The star's guts spew forth at 12,000 miles or more a second, creating shock waves whose Mach numbers soar into the thousands. The ensuing maelstrom creates elements both familiar (such as carbon, oxygen, and iron) and exotic (such as arsenic, rubidium, and krypton), fleshing out the top half of the periodic table.

Astrophysicists call that short-lived spectacle a supernova—specifically, a core-collapse supernova. During its first several weeks, it can outshine billions of suns. Nowadays, investigators continually identify new supernovae. Those discoveries come about not because of the explosion but because of the shock waves that pass through the star's own outer layers, rendering the event visible across millions and even billions of light-years.

If you think a supernova shock front is big and bad, picture what happens when an entire galaxy crashes into its neighbors.

Take the cluster of large galaxies called Stephan's Quintet, a collision of four galaxies plus one interloper that just happens to show up in the foreground of all images of the other four from Earth's perspective like a galactic photobomb. The colliding members of the main group have torn gas clouds from their host galaxies and strewn them hither and yon, making a real mess of the environs. One protagonist, plummeting toward its three neighbors at speeds exceeding Mach 100, has created a bow shock (a bow-shaped shock wave) so immense that its leading edge is larger than the entire expanse of our own Milky Way galaxy. Speaking of the Milky Way, it's falling toward the Andromeda galaxy. Shock waves are forecast for a few billion years from now.

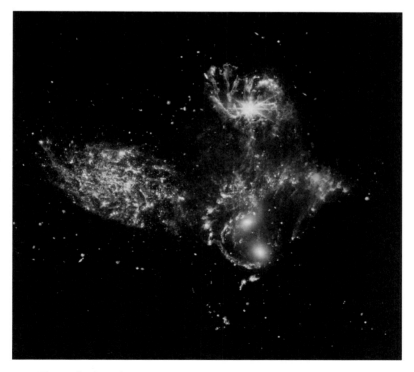

The mid-infrared instrument, a combined camera and spectrograph aboard the James Webb Space Telescope, shows never-before-seen details of Stephan's Quintet, a visual grouping of five galaxies, providing insight into galaxy evolution in the early universe.

When it comes to shock waves, neither atomic bombs nor hydrogen bombs nor movie explosions can compete with gamma-ray bursts (GRBs), the greatest blasts in the universe. Although not yet fully understood, GRBs may represent the death throes of a supermassive star under particular circumstances of rotation and environment, as well as perhaps a particular orientation to our line of sight. Regardless, they're bright enough to be seen from Earth with orbiting gamma-ray telescopes, no matter where in the universe they take place.

A cosmic gamma-ray burst is akin to a supernova on steroids (not to be confused with a supernova on asteroids, which isn't a thing). Between the burst and Earth lies the almost airless vacuum of space, and so there are gaps in the medium that might otherwise carry the burst's annihilating sound and fury down here to us. The resulting silence confirms that in space, not only will no one hear you scream, no one will hear you explode either.

A DARK MYSTERY

On this cosmic journey of science and discovery, pursuing the nature of space beyond our solar system, there's a lot going on across the vastness of space that doesn't meet the eye, or even the telescope. And once again, curious, befuddled scientists have tasked themselves with figuring it out.

Why is the sky dark at night? Seems like a silly question—as silly as it once was to question whether the Sun really does travel around Earth, or whether air is full of matter. The world appears a certain way, and so that's just how it is. The night sky is dark, and so why shouldn't it be?

But ponder this: If photons travel uninhibited through the vacuum of space, and if space is infinite, shouldn't the night sky

continued on page 202

THE REAL DEATH STARS

O ur galaxy hosts a couple of supernovae every century, which happen to be copious sources of gamma and x-ray radiation. If a star within 50 or so light-years of Earth goes supernova, it could cause serious damage to our atmosphere, perhaps jeopardizing most terrestrial life.

Scientists think that two and a half million years ago, one or more supernovae erupted near the solar system, sending a devastating maelstrom of radioactive particles and energy toward our planet. The luminosity of the explosion would have outshined the combined brilliance of more than a hundred billion stars.

The ensuing electromagnetic pummeling swept away swaths of Earth's protective ozone layer, while deadly doses of radioactive particles exposed large mammals and other organisms to cancer-causing DNA damage and likely caused (or at least kick-started) a massive extinction event considered the end of the Pliocene epoch, 2.6 million years ago. The 60-foot, 60-ton megalodon shark, occasionally resurrected as the antagonist of oceanic horror films, went extinct during this time, as did many other marine flora and fauna. A nearby supernova may have also caused the end-Devonian extinctions of nearly 400 million years ago—the era just before the Carboniferous period, which has given us all our fossil fuels.

Hundreds of stars reside 50 light-years or less away from Earth, the danger zone for supernovae that could cause total devastation to our planet. But rest easy. Astrophysicists agree that none of these stars pose any threat of going supernova. Our worries are better addressed by combating climate change, deflecting asteroids, and preparing our thousands of artificial satellites for electronic damage from solar flares.

Messier 1, the Crab Nebula, is the beautiful aftermath of a long-ago supernova.

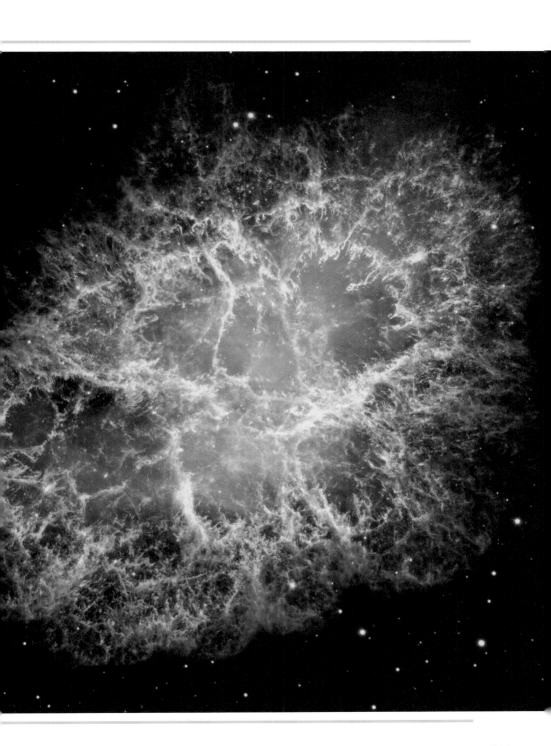

continued from page 199

be blindingly white, awash with the light of an infinitude of stars? Instead, it appears as a dark dome, punctuated here and there with tiny points of light. On a clear night in a darkened wilderness, the unaided eye can distinguish about 5,000 stars across the sky. In a typical suburb, that drops by a factor of 10. In bright and bustling cities, it drops by another factor of 10. We are now down by a factor of 100, leaving urban dwellers with no more than a few dozen stars, the Moon, and the Sun.

But let's say we take our telescope above the cities and Earth's distorting atmosphere. The Hubble Space Telescope, circling our planet in low Earth orbit, has gifted us a gorgeous gallery of the universe—some images capturing thousands of galaxies, each galaxy containing some hundred billion stars. But even Hubble's images show a mottling of dark patches amid a thicket of bright blots. Clearly, then, either the number of stars in the universe is not infinite, the universe itself is not infinite, or something else is going on. The answer, as is so often the case for cosmic conundrums, is both.

Imagine standing in a dense forest far from the trail. Everywhere you look, you see trees. Some are farther, some are closer, but no viewpoint offers a clear sight line to an unobscured horizon. The foliage of every tree blends with the branches of its neighbors until they lose any individual distinction. The color of your world is the collective color of the forest flora. If the universe is infinite, then in every direction we look, every pinprick-size spot in the night sky should end in starlight, and, like the trees converging in our forest, the starlight of each star should blend with starlight of its neighbors until we see only an amalgamation of illumination in every direction we look. And yet between each glistening star we see innumerable sight lines terminating in darkness. Why?

continued on page 206

STELLAR WARFARE

N early every space-based sci-fi film ever made devoured a large chunk of its budget with at least one climactic space explosion—and got it wrong nearly every time.

Among the most iconic and embarrassingly inaccurate space explosions ever depicted appeared in *Star Wars: Episode IV—A New Hope* (1977). The destruction of the Death Star generated a thunderous boom and an outward-radiating blast. A real ship, even one of such colossal proportions, would indeed explode in a short-lived, spherical blaze—but only as long as the ship's fire-feeding oxygen reserves remain available. Once introduced to the vacuum of space, the blast would quickly fizzle. More important, this bright, brief spectacle would play out in complete silence.

Star Wars, as the title implies, regales its fan base with many scenes of space warfare. The *Millennium Falcon* and the smaller TIE fighters dart left and right, up and down, banking turns and maneuvering as though they were airplanes supported by Earth's atmosphere and as though those instantaneous directional shifts weren't happening at thousands of miles an hour. In space, directional shift does not result from the simple turn of a steering wheel. To turn on a dime, a war-fighting spaceship needs a sophisticated set of directional nozzles fired at the correct angles and with sufficient thrust. To do an about-face, the ship would first need to undo its forward momentum by firing rockets in the opposite direction with enough energy to slow down and stop before reversing its path.

Let's say the TIE fighter pilots exploit some bleeding-edge technology that enables them to instantaneously switch directions. They'd best be buckled tight into heavily padded seats. Even so, their internal organs will be jostled around at speeds high enough to turn them into human milkshakes. Perhaps ship-to-ship stellar warfare is best left to drones and droids.

THE CREDIBLE HULK?

arvel's green monster might be the most scientifically implausible superhero ever conceived. While other superheroes might seem preposterous—such as the flying alien with x-ray eyes and a susceptibility to Kryptonite (Superman), or the handsome, hammer-wielding Norse god of thunder from the realm of Asgard (Thor)—their origin stories don't invoke science. The Incredible Hulk's, however, does.

Inspired by the classic tales of Frankenstein and Dr. Jekyll and Mr. Hyde, the legendary comic book creator Stan Lee spun a tale about a meek theoretical physicist named Bruce Banner, a Dr. Jekyll character. After an experimental bomb mishap bathed him in a potent dose of gamma radiation, his altered DNA enabled him to transform into a Mr. Hyde—a supersized, superstrong mutant with anger management issues.

Even a mild dose of gamma rays would have damaged Banner's DNA enough to send him the way of the megalodon within weeks. But let's write that off. When Banner becomes the Hulk, his skin turns green. Why not violet, the color closest on the visible light spectrum to gamma rays? Let's write that off, too, and call it artistic freedom. What we can't write off, however, is Hulk's trademark alteration: his taller, beefier body, capable of hurling a car as easily as a baseball.

Where's all that additional mass coming from? Einstein supplied us the recipe for gaining mass from surrounding energy: $E = mc^2$. But if Hulk creates the necessary quantity of mass from the energy of his surroundings, the very city he's trying to save would implode.

If the Hulk instead expanded without adding mass from elsewhere, he'd lose density for the same reason that ice cubes float. Liquid water expands as it freezes. It doesn't add molecules; it just takes up more space with the same material it had before. Your density is exactly equal to your mass divided by how much space you occupy (your volume): $d = m/v$. When ice expands, the denominator (in this case, v for volume) increases—which means that the overall value, the density, decreases. That's why an ice cube made of water floats on liquid water.

Without added mass, the fearsome Hulk might be as squishy as a marshmallow, but less cuddly. A single punch to his fluffy core would send him bopping down the street like a beach ball. A scientifically credible Hulk would make a lousy superhero.

Perhaps the most compelling unanswered question in Hulk's superhero story is, How do his pants stay on? They must be the stretchiest, strongest cargo pants in the universe. Now that's a material NASA would love to replicate.

Dr. Bruce Banner goes full Hulk in 2008's *The Incredible Hulk.*

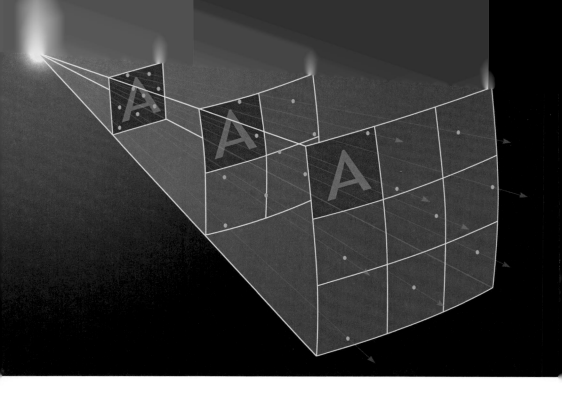

A schematic demonstrating how light intensity decreases over distances

continued from page 202

In the early 19th century, German physician and astronomer Heinrich Wilhelm Matthias Olbers proposed this very question, which came to be known as Olbers' paradox, though many others had pondered the problem long before he popularized it.

The solution is twofold. The first comes from the behavior of light over distances, and the second from the behavior of the universe itself. Let's start with the one that's easier to digest.

Two stars in our night sky may appear equally bright, but that doesn't mean they are of equal size or equally distant from us. A highly luminous distant star could certainly outshine a dim nearby star. In fact, most stars you see in the night sky are highly luminous at great distances. Sirius, the brightest star in our night sky, is smaller than Earth and 8.6 light-years away, while the next brightest star, Canopus, is larger than our Sun yet a staggering 310 light-years away.

Light dilutes over space in accordance with the inverse square law, the law that Isaac Newton initially set forth to describe gravity's influence over distance and that was later discovered to apply to light as well. The equation tells us that a light three times farther away than another light of the same luminosity does not appear three times as dim, although it would be perfectly reasonable for you to think so. No, it's nine times as dim. If a star is five times farther away, it appears 25 times dimmer. At great enough distances, we won't perceive the star at all.

Imagine loading up a slingshot with a fistful of pebbles. A target standing a foot away would get pummeled by every one of those pebbles. Move that target three feet away, however, and nine times fewer pebbles will hit it. Now suppose you have magic pebbles that will travel forever in whatever direction you initially launch them. As your target moves farther and farther away, its likelihood of taking a hit decreases exponentially, until the target arrives at a place where the possibility of intercepting a single pebble reduces to near zero.

On its own, the law of light dilution over distance cannot answer Olbers' paradox. If the universe is infinite, then even if luminosity decreases with distance, a single photon from a sufficient portion of the universe's infinitude of stars would yield a white night sky. Think of our magic-pebble slingshot. If you launched infinite pebbles from infinite slingshots, your target would take an onslaught of pebbles to every square inch of its body, no matter where it moved.

Now comes the second part of the solution. It dates back to the 1920s, when two scientists, working independently, didn't merely solve the paradox but also revolutionized astrophysics and dealt yet another blow to humanity's ever fragile, ever delusional ego.

In 1924, Edwin Hubble discovered that the Milky Way galaxy was not the entire universe, as many scientists had contended, but rather one among many "island universes"—a term the German philosopher Immanuel Kant proposed two centuries earlier. In 1927, after studying Einstein's new theory of relativity, Belgian physicist and Catholic priest Georges Lemaître proposed the idea of an expanding universe that could be traced back to a single point: the "primeval atom," which he later called "the beginning of the world." Today, we call it "the Big Bang." Lemaître's mentor, Arthur Eddington, didn't bother to read his paper until a few years later, and Einstein was not ready to embrace the notion of an expanding universe, so Lemaître's pivotal contribution was for a long time ignored.

Meanwhile, Hubble was busy studying the Doppler shift of his newly discovered island universes. An effect similar to the one demonstrated by a trainful of horn players in 1845 is a feature observed in all waves, including light. When invoked in discussions of cosmic light, it's called redshifting or blueshifting.

Recall that as the train approached the station, the pitch of the horns rose, because the sound waves shortened and their frequency increased. As the train passed the station, the trumpets dropped in pitch as the sound waves lengthened. On the visible light spectrum, shorter wavelengths are blue and violet, while the longer wavelengths are red. If the sound waves of the train correlated to visible light, we might see the color of light shift toward blue as the train approached and then shift toward red as it departed. Stars, galaxies, and everything else in the universe that emits light behave in a similar way. Waves coming from an approaching source of light and shortening are said to be blueshifting, and waves coming from a receding source of light and lengthening are said to be redshifting, whether or not their actual frequencies lie on the visible spectrum.

Hubble showed that the distant island universes are systematically rushing away from us, or redshifting—and that the

POE'S PREDICTION

n his 1848 book-length essay *Eureka: A Prose Poem,* the American writer Edgar Allan Poe, known for his macabre mystery tales, described a solution to Olbers' paradox decades before astronomers accumulated the data needed to fully understand the solution themselves:

> Were the succession of stars endless, then the background of the sky would present us a uniform luminosity, like that displayed by the Galaxy—since there could be absolutely no point, in all that background, at which would not exist a star. The only mode, therefore, in which, under such a state of affairs, we could comprehend the *voids* which our telescopes find in innumerable directions, would be by supposing the distance of the invisible background so immense that no ray from it has yet been able to reach us at all.

For a poet entirely untrained in physics or astronomy, Poe's statement was startlingly correct.

most distant galaxies are retreating faster than the nearer ones. In other words, his analysis was evidence of Lemaître's prediction that the universe is expanding. We finally realized that the universe had a beginning and that the universe must have a horizon, an edge, a limit. The observable universe is finite, and it is expanding.

Little did humans in the 1920s know that they were not done with the humiliation imposed by their own astrophysicists. Hubble documented that galaxies retreated in all directions. And so, we still appeared to be at the center of whatever the universe was. It was a scaled-up Ptolemaic worldview, soon to be torn asunder.

The galaxies are not rushing away from us. Rather, the interstitial fabric of space itself is expanding, carrying all the galaxies along for the ride. Think of it as poppy seeds embedded in the volume of a baking muffin. As the muffin expands, the distance between each poppy seed grows. For any single poppy seed, the appearance of all its poppy seed neighbors in retreat will appear to validate its feelings of being the center of the muffin. Victimized by the ultimate delusion of centrality, however, every other poppy seed feels exactly the same. Not only is the universe (or muffin) under no obligation to make sense to you; it's under no obligation to make you feel good about yourself.

The final solution to Olbers' paradox—the Big Bang theory coupled with the expanding universe and the finite speed of light—hit us with the latest of the great humblings to emerge from cosmic discovery: The universe is finite. And we are not at its center.

Yet the humblings hadn't finished. Another one, perhaps the greatest since Copernicus, was soon to come.

MEASURABLE YET UNIMAGINABLE MAGNITUDE

Half a century after Edwin Hubble's discoveries, the space telescope named in his honor informs our current estimates for the total number of galaxies in the universe. That number has reached several hundred billion and may soar as high as a trillion. Meanwhile, each of these galaxies contains about a hundred billion stars, on average. As our telescopes take us ever deeper into the universe, the sheer scale of objects and phenomena quickly becomes unimaginable to our feeble minds.

To grasp the magnitude of these discoveries, consider this apparently simple question:

What's the difference between a billion and a trillion?

Let's count seconds instead of stars. People who are 31 years old will have lived for a billion seconds a few months before their 32nd birthday. If you're not yet 31, it offers a good excuse to plan a party—but mark your calendar and prepare to take a superfast sip of champagne before the moment slips by. If you've already missed your one billionth second, there's always your two-billionth-second birthday to look forward to at age 63. Some lucky few will even live to see their three billionth second at age 95.

If you had a hundred billion dollar bills and a lot of time on your hands, you could lay them end to end to circle Earth 200 times. You could then vertically stack the remaining money to reach the Moon and back 10 times. The richest person in the world today could do that exercise twice.

A trillion, on the other hand, is larger by a factor of a thousand. A trillion seconds ago, the last Neanderthals inhabited Earth, and cave dwellers painted bison on their stone walls.

Now let's combine hundreds of billions with trillions. Multiply a hundred billion galaxies by a hundred billion stars per galaxy, and you get a billion trillion stars in the observable universe. That's a million times more stars than the total number of sounds and words ever uttered by all the humans who have ever lived.

But we're not done.

Astrophysicists now think that at least one planet orbits every star, on average. Welcome to the exoplanet: the next humbling of humanity, verified by scientific enquiry in 1995. That year, two

> If you had a hundred billion dollar bills and a lot of time on your hands, you could lay them end to end to circle Earth 200 times. You could then vertically stack the remaining money to reach the Moon and back 10 times.

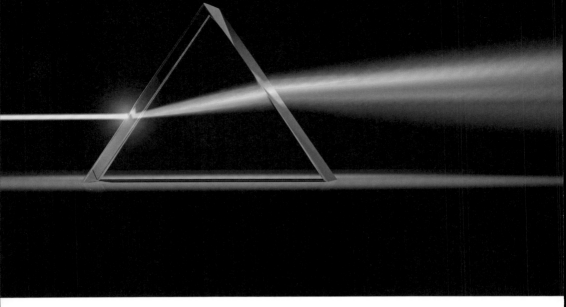

A prism splitting light into the color spectrum

Swiss astronomers, Michel Mayor and Didier Queloz, noticed something odd. Periodic shifts in the speed and position of a Sunlike star suggested that something unseen but real was in orbit around it, throwing off the star's center of gravity enough to cause it to wobble. It turned out to be a planet—an exoplanet— the first one found beyond our solar system around another Sunlike star. In 1992, three exoplanets were found orbiting the corpse of a star—a so-called neutron star—causing not a wobble but a change in its pulse. Before today's conventional naming formulas, these exoplanets received macabre monikers reminiscent of undead fantastical monsters: Draugr, Poltergeist, and Phobetor. Since Mayor and Queloz's discovery, nearly 4,000 star systems have been observed, containing more than 5,000 confirmed exoplanets.

When Isaac Newton split white light into the constituent colors of the rainbow, he planted the seeds of modern astrophysics. By the mid-1800s, the new science of spectroscopy would empower astrophysicists to know the temperature, rotation, movement, magnetic field, and chemical composition of stars. Exoplanet hunters today rely on both spectroscopy and powerful telescopes to detect a planet's presence and, in some cases, the composition of its atmosphere. Advances in telescopes and auxiliary technologies continually upend our assumptions about the universe and our place within it, by showing us the universe for what it is, rather than how it appears to our senses— and how it appeals to our fragile ego.

THE HIDDEN MESSAGES IN RAINBOWS

In the 20th century, we would learn that light from the most distant galaxies contains the chemical signature of countless otherwise invisible cosmic objects that it encounters on its way to our telescopes. The new means of analysis can be traced back to 1802, when English chemist William Wollaston improved upon Newton's prism and split the Sun's spectrum of colors even further, discovering dark bands invading the smooth continuum of colors.

Decades of experiments with splitting light eventually revealed that the dark bands arranged themselves according to the temperature of the light source, as well as the chemical elements interacting with that light. From these observations came an encyclopedic compendium of spectral "line" patterns, each representing the unique fingerprint of an element.

In 1868 two scientists, working separately, noticed a suspicious band in the Sun's spectrum unlike that of any known element. French astronomer Pierre Janssen and English astronomer Joseph Lockyer had discovered, for the first time, an element in space for which there was no known earthly counterpart. Lockyer named it helium—for Helios, the Greek god of the Sun. Although today you can find festive floating containers of the stuff at any children's birthday party, scientists didn't isolate it on Earth until 30 years after its discovery in the Sun. We now understand it to be the second most abundant element in the universe, after hydrogen, critical to the formation of all the heavier elements that make up everyone and everything you know and love, as well as everything you don't know and hate.

Although spectroscopy accurately and consistently revealed both known and unknown elements, nobody knew why. Eventually, spectroscopic experiments led Maxwell to formulate the equations from which Einstein derived his special theory of relativity. But it wasn't until 1913 that Danish physicist Niels Bohr provided a satisfactory understanding of how and why atoms generate spectral lines. He proposed that the electrons of every atom respond to light by jumping back and forth from one energy level to the next within the atom, causing the absence (or presence) of lines in an otherwise continuous spectrum of light. Bohr's atomic model helped set the stage for the modern era of quantum physics.

Fast-forward to the 1990s, when our exoplanet hunters, Mayor and Queloz, pioneered a new type of spectroscopy. Their method could accurately measure the subtle Doppler shifts of stars, caused not by their moving from one place to another in space but by their jiggling ever so slightly in response to the gravity of another object in orbit around them. Those subtle jiggles revealed the world's first exoplanet circling a Sunlike star, 51 Pegasi, in—where else?—the constellation Pegasus.

To account for the star's jiggling, the orbiting object—51 Pegasi b—had to be huge, half the size of Jupiter. But that wasn't the weirdest part of the discovery: One year on that giant gas

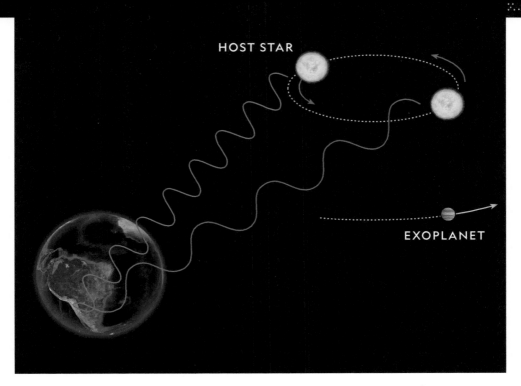

EXOPLANET

An illustration of Doppler spectroscopy: When a star moves toward
an observer, its spectrum is blueshifted; when it moves away
from the observer, it is redshifted.

world—one full orbit of its host star, in other words—turned out
to be a mere four days. Compare that with our Sun's nearest and
smallest planet, Mercury, which is teensier than Jupiter's largest
moon and whose year lasts 88 days.

Mayor and Queloz's discovery contradicted everything we
thought we knew about the formation of star systems. Although
multiple observations confirmed their finding, many research-
ers, stuck on the idea that our solar system must be representa-
tive of them all, refuted the planet hypothesis and offered
alternative explanations for the star's strange orbital behavior.
Although several more exoplanets of similar size and orbital
period were discovered using the same method, the skepticism
continued, in part because the discoveries did not derive from
direct observations of the planets. Their existence was inferred
entirely from the behavior of the host star.

Then, in 1999, one of these inferred planets, Osiris, was finally directly observed via the transit method—the same method Jeremiah Horrocks used to measure the distance between Earth and the Sun during the 1639 transit of Venus and thereby to estimate the size of the entire solar system. Astrophysicists predicted when Osiris would next pass in front of its host star as seen from Earth, and they were ready. As expected, the star's brightness dipped just a little during the planet's transit and returned to normal, all at the moment predicted.

The doors to exoplanet hunting flew open. An exercise once met with derision quickly became a booming, bustling field of study. The Sun was demoted from its role as the center of the only known star system to just one among untold others. Yet again, our universe had expanded before us.

> The doors to exoplanet hunting flew open. An exercise once met with derision quickly became a booming, bustling field of study . . . Yet again, our universe had expanded before us.

THE HUNT FOR EXO-EARTHS

Cosmic spectroscopy from below Earth's atmosphere is limited to observations of only those wavelengths of light that successfully penetrate the sea of atmospheric molecules above us—mostly visible light and longer-wavelength radio waves. The ozone layer blocks incoming ultraviolet (UV) light and x-rays, while atmospheric water molecules distort and absorb infrared and microwaves. What makes water a powerful greenhouse gas is also what complicates Earth-based astronomy.

If we want to observe celestial objects that emit these otherwise blocked bands of light, we must do so from above the

ALIEN FM

I f humans want to keep our doings secret from eavesdropping
aliens, we should communicate only via AM radio.

Earth's ionosphere is the atmospheric layer awash in free-
floating electrons ripped from their atoms by the high-frequency
ultraviolet, x-ray, and gamma radiation our Sun delivers. When the
longer, lower-frequency AM radio waves interact with those electrons, some or all of that energy is bounced back down to Earth's
surface, permitting communication far beyond your horizon sight
lines, especially at night. Under certain conditions, this back-and-forth
can continue across continents, allowing humans to "hear" one another
from great distances, the way the ocean propagates whale songs.

If you've ever listened to the radio on a long road trip, you already
know you can get some fuzzy AM radio from three states away, while
a crystal clear FM station will fade quickly from one side of a single
city to the other. The much higher frequencies used by FM radio
stations as well as satellite television don't benefit from an iono-
spheric ricochet, so FM frequencies can only reach the horizon, and
cannot follow Earth's curved surface. But that doesn't stop those
signals from escaping upward into space. Nearly every FM and TV
frequency transmitted by humanity—whether conveying *I Love Lucy*
reruns or multimillion-dollar Super Bowl halftime commercials—is
susceptible to interception by any alien with a sensitive radio
receiver.

clouds. Enter Hubble and other space telescopes. With their
help, hundreds of exoplanets were discovered between 1999 and
2009. But significantly upping the discovery rate of these tiny,
dark, moving specks required new kinds of telescopes, designed
specifically for this purpose.

A selection of the planetary discoveries made to date
by the Kepler space telescope

NASA's 2009 Kepler mission launched with a single directive: to hunt for Earthlike planets within a Sunlike star's habitable zone, also called the "Goldilocks zone." This is the orbital region around a star where a planet will find itself at just the right temperature—not too cold, not too hot—to sustain liquid water on its surface, a prerequisite for life exactly as we know it. Around highly luminous stars, the habitable zone is wider; around cool and weakly luminous stars, the habitable zone is narrow, hugging its host more tightly for needed warmth.

To the delight (and, perhaps in some cases, distress) of Earthlings, the Kepler mission detected droves of exoplanets.

Among the thousands of new worlds Kepler detected, hundreds were more or less the size of Earth with a similarly rocky composition. Several of those exo-Earths orbit within the host star's habitable zone. Statistical analysis of the mission data showed that not only do planets outnumber stars in our galaxy, but also that the Milky Way might harbor around 300 million Earthlike worlds within a habitable zone. Astrobiologists think at least one of these exo-Earths may very well reside within a mere 20 light-years from us, awaiting detection.

From the first moment anyone looked skyward, stars that guided and captivated us were more than stars all along. Over the centuries, it became clearer and clearer that space is not just a canopy of stars, nor is it some passive volume in which cosmic objects live and work. It's a vibrant, energy-filled place that requires us to first understand its influence on our observations before we can claim full knowledge of anything in the universe. Perhaps, as you read this, some as-yet-undiscovered life-forms are scratching their bald alien heads, wondering how much of their version of the cosmic tapestry they actually understand.

What next great humbling of humanity's hubris awaits us? Astrobiologists work tirelessly to seek out evidence that might topple Earth's position as the only life-friendly planet. With hundreds of millions of Earthlike exoplanets predicted to be roaming unobserved within just our own galaxy, life as we know it may be as ubiquitous as the once preposterous exoplanet.

Spectroscopy equips astrobiologists to search not only for Earthlike planets but also for signs of life, chemically inscribed in a planet's atmosphere: biosignatures, telltale molecular clues to life. All living things, as far as we understand them, undergo metabolic reactions to convert mass into energy within a liquid medium.

The waste products of these reactions are the smoking guns we seek. An alien using spectroscopy to look for Earth's biosignatures would see abundant oxygen in our lower atmosphere, issuing primarily from photosynthetic life, as well as ozone molecules in the stratosphere. Methane, produced primarily by

continued on page 222

EAU DE OUTER SPACE

You may have heard that space smells like various things: ISS astronauts and Moonwalkers swear it smells of gunpowder and charred steak, with a smidge of rotten eggs. But attempt to catch a whiff outside your space suit, and you'll asphyxiate.

Perhaps these astronauts are detecting the aromas of their own upper lips? Turns out, assorted stinky particles that adhere to their space suits and equipment while they're out for a spacewalk, perhaps combined with ozone produced when the space capsule repressurizes, result in a distinct Eau de Outer Space. But beyond low-Earth orbit and the Moon's surface, only spectroscopy can offer clues to the bouquet of the universe.

In 2009, a group of astronomers went searching for amino acids in a distant nebula. They detected, among many other complex molecules, ethyl formate—known on Earth to smell of rum, to taste of raspberries, and to be useful as a solvent and pesticide. Headline after headline soon proclaimed that outer space smells like raspberries.

Not likely. Even if we could scoop up enough assorted nebular molecules to fill a jar, bring it to Earth, and take a sniff, we would almost certainly fail to detect that scent. We might also poison ourselves with other toxic molecules in the brew. Furthermore, if we apply this logic to our own early solar nebula, we might be inclined to assert that it tastes like blood, owing to the high iron content. (While "Space Tastes Like Blood" might make a catchy headline, it's as misleading as the jests about raspberry daiquiris deep in the universe.)

But an astonishing spectroscopic discovery was buried in this silliness. The complex molecules found within that faraway nebula strongly suggest the possibility of amino acids—the building blocks of proteins and of life as we know it.

Astronaut Bruce McCandless II floats freely beside the space shuttle *Challenger* during the first ever untethered spacewalk.

continued from page 219

anaerobic bacteria, is another useful biosignature. Without constant replenishment, both would wane. Oxygen would be chemically reabsorbed back into Earth's surface, and methane would break apart and turn into carbon dioxide within a couple of decades, gifting its constituent hydrogen atoms to space.

By themselves, however, biosignatures do not confirm life. Most of these gases can be accounted for by other, albeit unlikely, nonbiological explanations. Although we could build a good case for exo-life, based on a planet's habitability and detected biosignatures, only an exploratory voyage can offer confirmation—or denial.

SPACE PILGRIMS

If we did find a new Terra—an Earthlike world with a breathable atmosphere and surface liquid water—the urge to explore it would outweigh even the urge to put colonies on Mars. If exo-Earths are detected within about 10 light-years' distance, their atmospheres could be spectroscopically analyzed for the presence of ingredients necessary to host humanity. If so, we might have an actual backup planet—a true Planet B, no terraforming required. The James Webb Space Telescope is exquisitely equipped to create a catalog of such candidates.

Ten light-years may not sound so terribly far. It's a mere 10-year trip for a photon. Unfortunately (or fortunately), humans are not photons. The fastest-moving human-made object, the Parker Solar Probe, which orbits and occasionally grazes the Sun, could travel from London to New York in about half a minute—or to the Moon and back in just over an hour and a half. Even at such blinding speeds, though, this probe will only ever achieve 0.064 percent (one-fifteenth of one percent) of the speed of light—hopelessly slow for any business outside the solar system. If the

TECHNOSIGNATURES

While life on an exoplanet's surface will likely influence the atmospheric chemistry, we presume the by-products of civilization will too. Before the industrial revolution, humans on Earth had no important influence on Earth's atmosphere. But over a shockingly brief period of time—on the scale of decades, not centuries—we have significantly increased the greenhouse gases CO_2 (carbon dioxide) and CH_4 (methane). Add to this an overall drop in breathable air quality from smog, belched forth from factory smokestacks and auto exhaust, and you have a chemical signature in the air.

But wait, there's more. For a spell there, we were depleting our otherwise stable ozone by the release of chlorofluorocarbons—chemicals commonly used as refrigerants and the primary propellant in aerosol cans that hold deodorant and hair spray—causing a sizable hole. We might call these atmospheric signs of industrial civilization spectroscopic "technosignatures." If aliens were to observe Earth the way we view exoplanets from afar, and they saw these abrupt disruptions to a stable oxygen-nitrogen atmosphere, they would surely conclude that, though life-forms must exist, there was no sign of *intelligent* life on Earth.

spacecraft could somehow dislodge itself from the Sun's gravitational grasp (which it can't), it would still take six thousand years to reach Alpha Centauri (at four light-years away, the nearest star system to the Sun). Pile on some humans, necessities, and the bare minimum of rocket fuel to get our pilgrim spacecraft going, and its speed will rapidly reduce by orders of magnitude. If our craft travels at the same speed as the International Space Station—about five miles a second—then a trip covering the distance of four light-years will take about 150,000 years.

What's a spacefaring species to do?

Holding aside any hope of time travel or wormholes, a generational spaceship—that is, a space ark capable of keeping successive generations of humans alive and on course—presents the only option for physically visiting distant worlds.

We could cut down the length of the journey by improving our speed and reducing our ambitions from 10 light-years away to one. Even so, we're still looking at navigating the open ocean of space for tens of thousands of years. Pioneer, Voyager, and New Horizons—humanity's bobbing corks, adrift in the interstellar seas—all executed multiple complicated gravity-assist maneuvers to escape the clutches of the Sun's ceaseless gravity. To launch a large station comfortable enough to support generations of harmonious humans across interstellar space, chemical rocket fuel won't cut it.

Early science fiction envisioned a world of flying cars and hoverboards powered by a source of unlimited energy. But they were wrong. Energy continues to be the limiting factor of most of our imagined technologies. Beholden to the unpredictable surge and fall of gas prices, power outages, and climate devastation, we remain dependent on nonrenewable resources. With such economic, environmental, and technological constraints, even a journey to Mars can feel unachievable.

Modern sci-fi authors have called upon nuclear power to fuel our species' interstellar adventuring. Harnessing the power of splitting or combining atomic nuclei could hypothetically cut down the duration of a journey by orders of magnitude. Fission splits heavy atomic nuclei, creating the devastating energy

> A generational spaceship—that is, a space ark capable of keeping successive generations of humans alive and on course—presents the only option for physically visiting distant worlds.

found in atomic bombs. Fusion, on the other hand, combines atoms with other nuclei, rather than splitting them.

The Sun constantly undergoes fusion in its core: Under extreme pressure, two atomic nuclei merge, producing heavier elements and releasing enormous amounts of energy as a by-product. Fusion is four times more efficient than fission and four million times more efficient than burning fossil fuels. A multistage rocket propelled by nuclear fusion (which may one day drive our nuclear power plants as well) might get humans up to a whopping 10 percent of light speed or even better.

We know that the unyielding tyranny of the rocket equation, which decrees that you must burn more fuel to carry the weight of the fuel you have yet to burn (see pages 72–76), presents the first obstacle to human space travel, whether to the Moon, Mars, or beyond. But let's pretend we've jumped that hurdle and invented a ship big enough, with an engine powerful and efficient enough, to comfortably cart humans between planets and

A fusion reactor, or magnetic reconnection plasma thruster,
concept developed by the U.S. Department of Energy's
Princeton Plasma Physics Laboratory (PPPL)

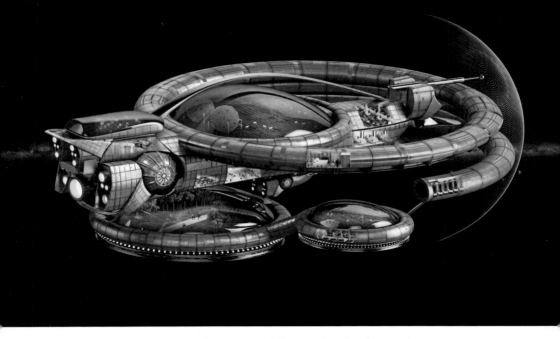

A conceptual generational ship, outfitted with everything
humanity would need to trek deep into the cosmos

stars on multiyear, multigenerational journeys. What hindrances might these spacefarers face?

An interstellar ark represents the ultimate conquest of nearly every hardship Earthlings collectively face today, ranging from a self-sustainable, equitable food system to psychological well-being. Even a Martian colony would struggle to become truly self-sufficient. Completely cutting ties with Earth would be an extraordinary feat. All material must come from somewhere, and spaceships cannot mine for new resources. Everything would have to be reused or replenished from a single supply intended to last for generations. The metals, plastics, and wiring that held your habitat together would eventually demand repairs. The shirt on your back would last only so many years before the threads ran bare. The medicine and food that sustained you would have derived directly and indirectly from Earth's living systems.

The most complex challenges, however, would be nonmaterial. What political system would ensure peace and justice among all inhabitants? Could entire generations of people live out their lives only to die without seeing the fruits of their labor—and knowing that their children won't see them either? Would the pursuit of Planet B for the sake of an endangered species' curiosity suffice? And is it even ethical to consider birthing a generation that would live and die within a society cut off from their home planet, with no choice but to endure?

But aren't these the very questions we all ponder down here on Earth? If we could solve all the problems that the society of a generational spaceship might face, then we needn't set out for a Planet B in the first place. As far as we can tell at present, Earth is an isolated world, alone in the universe, caught in orbit around our star for billions of years with no food, medicine, or air to replenish what we use or destroy. If we fail on this world, where we surely know our failure could mean extermination, can we reasonably expect different results out there, lost to the sea of space?

You might be wondering why we don't just send an armada of robots to explore other worlds in our stead. After all, they can travel unburdened by mortal needs or emotions, without the haul of supplies required to meet those needs. The answer is that, at least for now, humans are better at science. Even if AI manage to become our overlords, the job of scientist (and maybe that of comedian) might be one of the last humans still occupy.

Even after decades of exploration of the Martian surface by five rovers and one helicopter, each one equipped with multiple experiments and tools to survey their surroundings, we still cannot rule out the possibility that alien life is hiding somewhere on the red planet. Humans, by contrast, could accomplish in seconds or minutes what a rover accomplishes in days, weeks, or never. We can solve problems creatively, via a level of mental dexterity that robots can only dream of. And they probably don't dream.

continued on page 230

DEATH BY VACUUM

What would really happen if a human stepped into outer space without a space suit? In the film *Mission to Mars* (2000), Commander Woody Blake and his crew are forced to evacuate their spacecraft. In the commotion, Commander Woody is jettisoned from the ship on a hopeless trajectory into the dark unknown. To prevent his wife from risking her own life to retrieve him, he pops off his helmet, exposing his face, head, and body to the unforgiving vacuum of space. Seconds later, we see an undeniably dead Woody, his face and eyeballs frozen solid. It is a heroic scene, equal parts gruesome and devastating. In reality, he would indeed die—but not from flash freezing. Let's analyze the realistic outcome of a death-by-vacuum:

1. Burst Lungs and Asphyxiation

If Woody had held his breath before removing his helmet, the first thing he would feel is his lungs bursting and rupturing, because all the gas in his body, including in his bowels, would immediately seek out the region of less density surrounding him.

But let's say our commander exhaled completely before removing his helmet. Because our brains are replenished by oxygenated blood from our lungs, after 15 seconds, Woody's body would have used up all the remaining oxygen stored in his blood, and the deoxygenated blood heading for his brain would render him unconscious. But he still wouldn't be dead, and definitely not yet frozen.

2. Bubbling Blood

Lower-pressure environments yield much lower boiling temperatures, as we saw with the triple point of water thought experiment on Mars (see page 130). In the vacuum of space, the gases in our blood will begin to boil, causing our bodies to swell—though not to the point of our eyes bursting out of our faces, as seen in *Total Recall*. Saliva would therefore boil right off Commander Woody's tongue, and his eye capillaries would start to rupture. After a few minutes, he would likely die from lack of oxygen.

3. Sunburn

Without a helmet, Woody's face would not be protected from sunlight. Unless he finds himself in a heavily shadowed region, like behind a ship or planet, his skin would burn, not from heat but from high doses of the Sun's biologically hostile ultraviolet rays.

4. Freezing

The very last thing that would happen to Woody's corpse is freezing, but only if it weren't too close to a source of heat energy, such as a star. Over one or two days, the commander's body would freeze solid. The now perfectly preserved, sunburned, bloodshot, and bloated body would remain in space, slowly surrendering to the force of gravity of whatever object was nearest when he decided to remove his helmet.

One of the most iconic science fiction films of all time is Stanley Kubrick's *2001: A Space Odyssey,* released in 1968 and based on earlier stories by Arthur C. Clarke. *2001,* as compared with *Mission to Mars,* depicted a far more accurate reality of what could happen to an unprotected human body in space. The villainous artificial intelligence (AI) program, named HAL 9000 (Heuristically programmed ALgorithmic computer), attempts to lock the protagonist out of the spaceship, forcing him to jump into the unpressurized air lock without a helmet and engage the air pressure lever, all while exposed to the vacuum of space. The scene lasts exactly 14 seconds (in perfectly accurate total silence). That's just enough time for our protagonist to elude unconsciousness and save himself.

Of course, he might need more time to recover than the film allowed, but he did survive and did not immediately freeze, explode, or vaporize.

continued from page 227

But even if we can advance robotics to create the perfect android—something that can replicate human abilities and mimic creative thought—the urge to explore remains firmly embedded, as it always has, in the human psyche. If robotic exploration could satisfy that urge, then why are we determined to set up colonies on the Moon and Mars? Why do we continue to pour billions of dollars into newer and more powerful telescopes that might help us see a wee bit farther than before? Why are we driven to expand the perimeter of our human ignorance to the edge of the universe and beyond?

Edwin Hubble captured the sentiment within his 1936 book *The Realm of the Nebulae:*

> The explorations of space end on a note of uncertainty. And necessarily so . . . We know our immediate neighborhood rather intimately. With increasing distance our knowledge fades and fades rapidly. Eventually, we reach the dim boundary—the utmost

A cosmic sea of swirling gas and dust within the Large Magellanic Cloud, one of the Milky Way's satellite galaxies, as seen by the Hubble Space Telescope

limits of our telescopes. There, we measure shadows, and we search among ghostly errors of measurement for landmarks that are scarcely more substantial.

A necessary consequence of exploration and discovery is indeed the ever widening perimeter of ignorance that separates what we know from what is yet to be discovered. We're currently deep into our universe—its variegated contents and the not-so-empty space throughout—yet what's just outside the front door is a wild west of wonder that includes, among other oddities, questions of whether we inhabit a simulated universe and whether our universe is but one among an infinite number of others in a multiverse. Conundrums continue to multiply, and the frontier knows no bounds.

The cosmic journey continues.

PART 4

TO INFINITY AND BEYOND

"Never in all the history of science has there been a period when new theories and hypotheses arose, flourished, and were abandoned in so quick succession as in the last fifteen or twenty years."
—Willem de Sitter, 1932

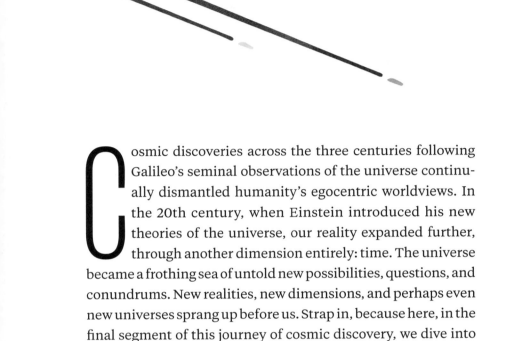

Cosmic discoveries across the three centuries following Galileo's seminal observations of the universe continually dismantled humanity's egocentric worldviews. In the 20th century, when Einstein introduced his new theories of the universe, our reality expanded further, through another dimension entirely: time. The universe became a frothing sea of untold new possibilities, questions, and conundrums. New realities, new dimensions, and perhaps even new universes sprang up before us. Strap in, because here, in the final segment of this journey of cosmic discovery, we dive into the eddies of black holes, where space and time warp beyond recognition. We travel into the past and into the future; we move at speeds faster than light; and we recognize, as far as our human awareness can carry us, what it means to travel to infinity and beyond.

From relativity emerged a cosmos that twists, bends, and ripples like a fabric, where gravity and speed alter the flow of time. But relativity also suggested a cosmos in constant motion—with a clear beginning but no clear end. These implications were not immediately apparent, nor were they accepted,

The formation of young stars in the breathtaking Pillars of Creation, as seen through the James Webb Space Telescope's near-infrared-light view

PREVIOUS PAGES: A moment of brilliant creation: the Big Bang, conceptualized in striking reds and golds especially by Einstein himself. It seemed so obvious that the universe was a constant, unchanging, forever entity that, even in the face of mathematical suggestions to the contrary, Einstein invoked a term in his equations that allowed them to match up with the universe whose nature he'd already presumed.

He called it the cosmological constant. We might think of it as a form of confirmation bias. It came to be known instead as Einstein's biggest blunder.

Long before Edwin Hubble and Georges Lemaître showed that the universe is expanding—which implied it had a beginning, now called the Big Bang—many of the world's cultures already embraced the idea that the universe underwent a moment of creation. Their religious texts say so. Nearly every faith offers a unique creation myth. The Hebrew Bible appropriately begins with the words "In the beginning . . ." According to the texts of the three Abrahamic religions (Judaism, Christianity, and Islam), heaven and Earth took six days to create. In the Buddhist and Jain religions, the universe undergoes eternal cycles of creation and destruction. Hinduism, which teaches that this current world cycled into existence some four billion years ago, may be the only known religion that conceives of the universe on timescales compatible with modern cosmology.

Perhaps if Einstein had been Hindu, he would have more easily accepted a nonstatic universe, one that could begin and end and shift along the way. As far as historians can tell, though, Einstein's thinking was unconstrained and unmotivated by the tenets of religious texts.

During the 19th century, while biologists, geologists, astronomers, and theologians were still squabbling over Earth's age, their debate was decoupled from thoughts of cosmic origins. Even if Earth had a specific birthday, it was contended, nothing could come of nothing. The universe must have always been here—an infinite, ageless place where humans and stars and

Some 14 billion years have passed since the Big Bang
(at the far left) catapulted the known universe into existence,
illustrated here from inception to present day.

planets and all things are born. No evidence suggested otherwise. Even to think such a thought was a fever dream, a fantasy
detached from sensible or philosophical reason.

With the notion of a static universe firmly anchored in his
philosophical substrate, Einstein developed his general theory
of relativity. He then applied it to the universe as a whole. His
1917 paper "Cosmological Considerations in the General Theory
of Relativity" embodies his assumption of a stagnant universe,
even though his own equations reveal the unsettling truth of an
unstable universe. Like a ball balanced at the top of a hill, ready
to fall in one direction or the other, the entire universe was either

continued on page 240

THE AGE OF EARTH

I n the 17th century, when the scientific revolution was in its infancy, the holy books still stood as the ultimate source of truth for Jewish and Christian believers. Their theologians' preferred method for deriving the age of Earth was to tally the "begats"—the ancient genealogies—recorded in the Old Testament. One 17th-century Irish cleric, James Ussher, having added up the biblical birthdays and life spans, arrived at a precise creation date of October 23, 4004 B.C.E. Others put it a millennium or so earlier. Six or seven thousand years must have seemed a sufficiently long time for all oral history to have occurred and all the planets in the sky to have formed. Back then, no Christian thinkers thought to wonder if the universe could be millions, let alone billions, of years old.

Two centuries later, the emerging fields of geology and biology would unveil an Earth, and therefore a universe, requiring a timescale much longer than previously imagined. The popular geological, as opposed to religious, view was that Earth was infinitely old—that Earth has, as 18th-century Scottish geologist James Hutton concluded, "no vestige of a beginning, no prospect of an end."

By the late 1840s, British physicist William Thomson, who would become the renowned Lord Kelvin five decades later, had established basic laws of thermodynamics, addressing the behavior of heat and the movement of energy from one place or one system to another. Based on his own calculations derived from the Sun's mass, surface temperature, and total energy output—and unaware there was a thermonuclear furnace at its core—he presumed it to be a slowly cooling, contracting ball of gas and pegged its minimum age at 20 million years. And since Earth could not be older than the Sun, Earth had to be the same age or younger.

These millions of years represented an irksome span of time: far too long to please religious fundamentalists, far too short to align with geologic data. To stir the pot further, in 1859, Charles Darwin would publish *On the Origin of Species,* which argued for a slow progression of organisms by natural selection. This idea, coupled with geological observations, now demanded an older Earth. Debates among physicists, geologists, and biologists raged for decades. Given that physics is the cockiest of the sciences and Lord Kelvin was that era's shining star, his opinion won out. Even American author Mark Twain weighed in on the estimate. "As Lord Kelvin is the highest authority in science now living," he said, "I think we must yield to him and accept his view."

By the late 19th century, Polish-born French physicist and chemist Marie Curie (below right, circa 1920) would co-discover radioactivity, which birthed nuclear physics and the cottage industry of radioactive dating. The age of certain unstable elements contained within rocks and fossils could be determined by looking at what fraction of them

had decayed to another element. This dramatically ratcheted up geological dating, yielding an age for Earth, and presumably the Sun, of about 4.5 billion years. We now know that, via thermonuclear fusion, the Sun can "burn" for billions of years and shall continue to burn for billions more. So let's thank Marie Curie, the only person to have ever won a Nobel Prize in two different scientific fields, for fueling and ultimately ending that great debate.

continued from page 237

expanding or contracting. Either option would lead to ultimate demise—a thoroughly untenable outcome, considering everything physicists actively or passively thought at that time. And so, Einstein inserted an antigravity term in his equations—a cosmological doorstop—to balance things. Behold an unchanging, ageless universe.

Five years later, in 1922, Russian mathematician Aleksandr Friedmann proposed that Einstein should consider a universe in motion. Einstein initially dismissed the idea. But Friedmann persevered, writing to Einstein directly, "Considering that the possible existence of a non-stationary world has a certain interest, I will allow myself to present to you here the calculations I have made." Within a few months, in the journal that had published those calculations, Einstein relented slightly. He admitted that, though Friedmann's calculations still allowed for Einstein's static universe, a dynamic universe might also be possible, writing: "I consider that Mr Friedmann's results are correct and shed new light."

Five years after that, Lemaître, working independently of Friedmann, published his own derivations for an expanding universe. Einstein rebuffed it, allegedly calling his physics "abominable." In 1929, Edwin Hubble collated existing observational evidence and concluded that galaxies were receding from Earth, and the more distant they were, the faster they were receding. Finally, in 1931, Einstein fully conceded. "The redshift of distant nebulae has smashed my old construction like a hammer blow," he said. That's how a highly literate scientist admits being wrong.

Out of this new cosmos arose new questions: If the universe is expanding, then it was smaller yesterday than it is today. Might the universe have begun as a singular explosion? At a place and at a time? If so, how old is it? How big is it?

WHEN THE BIG BANG WAS A BIG JOKE

Two things were clear to the English astrophysicist Fred Hoyle in the 1940s: The universe was expanding, and the universe cannot be younger than Earth. An entire universe, he reasoned, could not possibly have been created out of nothing. To reconcile all these presumptions, he championed a brand-new idea called the steady-state model, in which the expanding universe is infinite, ageless, and has always looked the same on average. How can anything perpetually expand yet always look the same?

To resolve this, Hoyle hypothesized that matter in the steady-state universe is spontaneously, continuously, and homogeneously created, and the energy of that creation is pushing galaxies apart. The spontaneously created matter slowly coalesces, making new stars and galaxies. This is how his statistical steady state of all objects and phenomena is achieved across the universe and through infinite time. Although Hoyle never explained the method by which matter appeared, he found the idea much more plausible than, as he mockingly described it in a 1949 BBC radio broadcast, "the hypothesis that all matter of the universe was created in one big bang at a particular time in the remote past."

Hoyle's sarcastic nickname for his opponents' theory stuck—and that's how the Big Bang got its name.

The precise velocity at which galaxies fly away from one another, as first measured by Hubble and subsequently fine-tuned by others, is called the Hubble constant (represented as H_0). If we assume the universe has always maintained the same expansion velocity, then we can throw it into reverse conceptually and with that measurement predict approximately how long ago everything must have been packed inside one little place.

To determine the age of our universe, all you need is the value of the Hubble constant—if in fact it is a constant—and some known distances to the other galaxies. A bit of back-of-a-napkin math will give you the age of the universe. With those data, Edwin Hubble retraced the steps of the universe back to a single point less than two billion years ago.

> All you need is the value of the Hubble constant . . . and some known distances to the other galaxies. A bit of back-of-a-napkin math will give you the age of the universe.

But by that time, geologists required that Earth be at least three billion years old to account for what they found in the rocks. Clearly, somebody was wrong. Hubble's two-billion-year-old universe elicited scorn from the geological community, dredging up memories of the age-of-Earth smackdown from decades past. Scientists knew to higher accuracy what Earth's crust told us, but then what of the expanding universe? The baffling conundrum of the age of the cosmos persisted a few more decades, eliciting in the meantime interesting new questions and conjectures.

In 1952, German astronomer Walter Baade, working in the United States, helped reduce some friction between the astro and geo camps when he deciphered a riddle hidden in a certain type of star. The discovery doubled the size of the universe, along with its age.

Hubble had calculated distances to nearby galaxies with a method called Leavitt's law, a cosmic yardstick the American astrophysicist Henrietta Leavitt invented during her work as a human computer at Harvard in the early 1910s. Leavitt was assigned to catalog variable stars, a type whose brightness varies over time. Among the stars of the constellation Cepheus, she found a category of variable star, now known as Cepheid variables, in which their pattern of brightening and dimming was directly related to their luminosity (in other words, its intrinsic brightness or absolute magnitude). This discovery allows you to infer the luminosity of variable stars by timing their cycle. Once you know the luminosity of the thing you're looking at, you measure its apparent brightness (in other words, how bright it looks from where you sit), apply a simple algebraic equation— and out pops the distance to the object. With this formula, Leavitt discovered how to estimate the distance from Earth to many objects within the Milky Way and even other galaxies.

For Edwin Hubble, this simple, brilliant method was a major key to unlocking millions of light-years and the distant galaxies they contained. But after a decade of careful observations, Walter Baade established the existence of not just one but two types of Cepheid variables, and that the second type placed the distant galaxies twice as far from the Milky Way as previously thought. Now, with the doubling of distances, the age of the universe also doubled—to nearly four billion years. The updated number would appease geologists and astrophysicists alike.

But we weren't done refining.

A few years later, American astrophysicist Allan Sandage, an assistant to Hubble and student of Baade, increased the age of the universe to 5.5 billion years when he realized that some of the brightest measured galaxies were not galaxies at all, but rather clouds of hydrogen. After Hubble's death in 1953, Sandage continued to refine Hubble's constant and the age of the universe to about 10 billion—much closer to our modern estimate of 13.8 billion years.

All the while, Fred Hoyle never renounced his ageless, steady-state universe, a model incompatible with Lemaître's Big Bang. For decades, the two views remained hotly contested. Alas,

Lemaître died two years before a fortuitous discovery cemented his theory.

In 1964, quite by accident, American physicists Arno Penzias and Robert Wilson took the temperature of the cosmos. While working at Bell Telephone Laboratories in New Jersey, the two were tasked to detect and weed out any microwave emissions that might interfere with our nascent communication satellites. Their equipment picked up a faint, puzzling garble that emanated from every direction of the sky, every hour of every day and night. Over the course of months, as Earth made its way around the Sun, the garble persisted. That signal, we now know, is the most ancient thing we've ever detected in our universe—the first light ever emitted.

When our universe was but a cosmological infant a billionth its current volume, 380,000 years after the Big Bang, light gurgled forth from its previous imprisonment within the broiling, roiling plasma of Everything. The escaping light left behind a pattern, inscribing all of space with a measurable artifact of its once seething brilliance. As space expanded, the wavelengths of that light stretched—redshifted—into the microwave part of the spectrum, imbuing all of space with a temperature of about three degrees K (that's three degrees above absolute zero on the Kelvin scale), or minus 454.76°F, detectable in every direction. We call this the cosmic microwave background, affectionately known as the CMB.

When we refer to the temperature of the universe, we mean the temperature of the cosmic microwave background. If we could venture to the darkest, coldest, emptiest void in the cosmos and stick a sensitive thermometer out our spaceship's window, that thermometer would read about 3 kelvins—three degrees above nature's lower limit of coldness. Just as we will never find

The brightness of RS Puppis, a Cepheid variable star here photographed by the Hubble Space Telescope, varies every 40 or so days.

a corner of the universe that is truly empty, we will never find a corner that does not feel these few degrees of temperature.

When light from a star expands outward as it journeys through space, the rays dilute rapidly and become undetectable over great distances. The CMB, on the other hand, pervades every square inch, and all of it has almost the exact same temperature, within a fraction of a degree. This factoid can mean only one thing: The universe was all in the same place experiencing the same event at the same time. Yes, credit the Big Bang.

When the CMB first escaped all those billions of years ago, it was about 3,000 kelvins and would have been detected (if anybody was looking) as mostly visible and infrared light. The space through which it traveled kept stretching. Now a thousand times bigger in every direction, the wavelengths have stretched by a factor of one thousand—to the long and cold yet detectable microwaves Penzias and Wilson picked up in 1964. Many billions of years in the future, they will redshift even further, into radio waves—a cosmic birthmark fading as it stretches but never quite disappearing against the skin of spacetime.

Penzias and Wilson were tasked simply to improve satellite communication on Earth. Yet in doing so, they stumbled upon the Big Bang's smoking gun: a discovery that would earn them a Nobel Prize. The static, steady-state model of the universe now received the final nail in its coffin. In its place, the dynamic universe imagined by Friedmann, Hubble, and Lemaître emerged triumphant and inevitable.

But from the glare of every cosmological discovery emerge new conundrums.

TO THE EDGE

We know the universe is expanding, and that it expanded from a single point. Furthermore, everything in the universe must

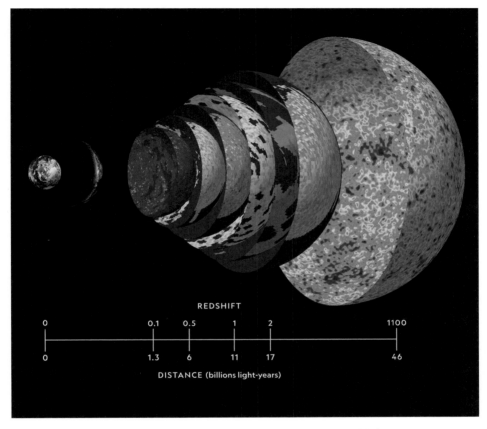

REDSHIFT

| 0 | | 0.1 | 0.5 | 1 | 2 | | 1100 |

| 0 | | 1.3 | 6 | 11 | 17 | | 46 |

DISTANCE (billions light-years)

CMB (cosmic microwave background) at varying redshifts

obey a speed limit. Reason would dictate that the universe must therefore have an edge—the ultimate defining barrier between space and, well, not space. However, reasonable though that might sound, it arises from a faulty premise, driven in part by the human ego. When we speak of the edge of the universe, we really mean the edge of the observable universe—the horizon beyond which light, constrained to a finite speed, has not yet had time to reach our telescopes. Spacetime expands like a poppy seed muffin, as already mentioned. The space between the poppy seeds expands, not the poppy seeds themselves.

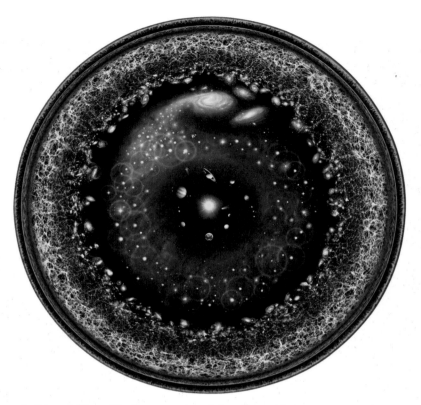

In this radial logarithmic conception of the observable universe, our solar system sits at the center. Expanding outward are the inner and outer planets, Kuiper belt, Oort cloud, Alpha Centauri, Perseus Arm, Milky Way galaxy, Andromeda galaxy, nearby galaxies, cosmic web, and cosmic microwave radiation, with the Big Bang at the outer edge.

Such comparisons are problematic, though, as we know that poppy seed muffins exist only within space—in our hand or in the oven or in our stomach. Instead, we could think of the muffin as Everything with a capital E. But if so, you might ask, since interstitial space is expanding, why isn't it expanding between stars, planets, or even between the molecules on Earth? Shouldn't the constellations we see in the night sky also lose their shape as the spaces between their components swell?

Perhaps we would encounter that in Hoyle's steady-state universe. But in our universe, gravity—often thought of as a weak force—is not so weak after all. The cumulated pull of matter keeps each galaxy—and each planet, star, and molecule residing within it—as a single tidy unit. They are the poppy seeds. Outside these tightly bound regions, space succumbs to the force of expansion, growing and thinning along the way.

If we now recast the question, we might ask what lies beyond the edge of the observable universe. In other words, where does Everything end? Space and time began everywhere at once. Everything was just smaller.

As far as we can deduce, what lies beyond the edge of the observable universe is more universe—no different from what we already see and know. More galaxies, more stars, more planets, more black holes. The total universe may be trillions of light-years across or perhaps infinite.

What about at the edge of even that? Beyond that edge of all edges, that uncharted territory of our cosmic map, we just don't know. We cannot know. Even if we could travel at light speed in pursuit of the fleeing galaxies, we could never catch up. Never.

SPACE/TIME

We can witness the past—not only through nostalgic photographs and videotapes, but with every glance at the stars. We see the Sun not as it is currently but as it was 500 seconds ago, because it takes that long for light to travel the distance between the Sun and Earth. If it instantaneously went cold and dark by the flip of a switch, we wouldn't know until 500 seconds later. If the same thing happened to the brightest star in the night sky, Sirius, we wouldn't know till nearly nine years later.

Imagine an advanced alien species on a planet a few hundred light-years away. If they focus a powerful telescope on Earth, they might see a blue planet with liquid water and an oxygen-nitrogen atmosphere. If that advanced species instantly teleports themselves here to take another look, they'll find a planet a few hundred years older than the version they first saw. This planet, they now realize, is overrun by a species consuming all the planet's resources at a rate sufficient to render it uninhabitable in short order. And that's just a few hundred light-years. Unaided by telescopes, we can see stars thousands of light-years away. What might have happened to them since their starlight reached our eyes? We cannot know. Not yet.

If we could teleport ourselves to a planet in a galaxy a billion light-years away, we might arrive in the middle of an exploding host star, or in the neighborhood of a mere wisp of a world long ago destroyed by an assailing asteroid. Observing deep space is not quite time travel or magic, but it's the closest thing we have to a crystal ball to the past.

If you're still not convinced that time and space are fundamentally entangled, then you're clearly not a professional event planner.

An effective party invitation must answer certain questions:

1. Where is the venue? In other words, what are the x and y coordinates on Earth's surface?
2. On which floor is the party? Or, what is the coordinate in the z dimension?
3. What is the date and time? Or, what is the fourth dimension coordinate?
4. Finally, how is the DJ?

Although Einstein posited the equations that would unite space and time, his former professor first fused these two otherwise separated concepts. Three years after Einstein published his special theory of relativity, German physicist Hermann

This graphic represents Minkowski's idea of worldlines in spacetime. You, the observer, are at the junction between the cones: the past (bottom) and the future (top). The plane (blue) is the hypersurface of present time.

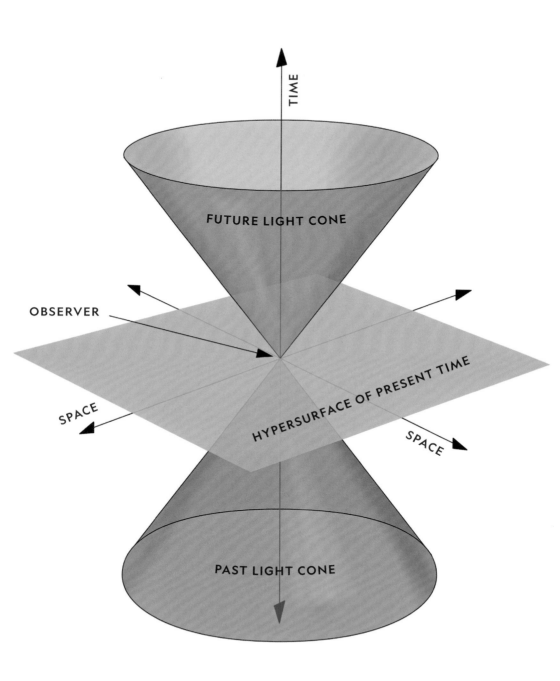

TIME

FUTURE LIGHT CONE

OBSERVER

HYPERSURFACE OF PRESENT TIME

SPACE

SPACE

PAST LIGHT CONE

Minkowski famously proclaimed an idea, hidden in plain sight, that our partygoers deduced: "Henceforth, space by itself, and time by itself, are doomed to fade away into mere shadows, and only a kind of union of the two will preserve an independent reality."

In that same speech, Minkowski introduced the term "world-lines." A worldline is the mapped trajectory, including the time coordinate, of any object, whether a particle or a person. A party happens because the worldlines of all the attendees intersect. What that simply, but profoundly, means is that everybody managed to occupy the same place at the same time.

Time and space are woven into the same fabric and are plotted together on worldline maps. Yet they are not the same, nor are they equally accessible or equally inevitable. We can choose to turn left or right, jump up or down. We can fly across oceans and travel to the Moon. But anyone who bears grief and regrets the past, anyone who prickles with anxiety for the future, anyone who finds it difficult to fully experience the present moment—which includes nearly everyone—knows that time remains an untraversable dimension.

> Time and space are woven into the same fabric and are plotted together on worldline maps. Yet they are not the same, nor are they equally accessible or equally inevitable.

We all move forward into our own future at exactly one second per second—but we cannot visit with the dead or meet our unborn great-great-grandchildren. Every passing second is at once a slammed and an opened door as time carries us away from what has been to what is and what will be until the day we arrive at our final spacetime destination.

Our regrets and worries may subside as our bodies weaken and age, but our pasts and futures remain as inaccessible to us as the farthest star in the sky—at least for now. Most worldlines are simply unreachable, constrained by our inability to maneuver between all four coordinates. Certainly the edge of the universe lies eternally beyond our worldline. You can no more visit those most distant galaxies than you can read this sentence again for the first time.

The clues to opening up our worldlines—unlocking pathways to the past or future—are not heard in the chants of mediums or discerned in the cards of fortune-tellers. They're written in mathematical equations. Einstein's theories predict and describe what the American physicist Kip Thorne calls "the warped side of the universe." This realm, he explains, contains things and phenomena formed from warped spacetime rather than from regular matter.

Thorne is perhaps best known for his Nobel Prize–winning work with the Laser Interferometer Gravitational-Wave Observatory (LIGO) team, which detected gravitational waves in 2015, just days after the observatory's advanced detectors were switched on—and exactly a hundred years after Einstein published the equations predicting them. The violent collision of two massive objects, such as black holes or neutron stars, sends out tiny ripples through spacetime at the speed of light. The displacement caused by a gravitational wave, measured by LIGO, is 10,000 times smaller than a proton.

LIGO's scientists envision using gravitational waves to observe the universe in somewhat the same way we already use electromagnetic waves to observe regular matter. Their facility has opened up a new window to the universe, as did Galileo's telescope four centuries earlier. Gravitational waves may reveal aspects of the warped universe that we've only just begun to grasp. And because they are not made of light, we may be able to detect gravitational waves dating back to an earlier time in the universe than the CMB.

TIME TRAVEL: HEADING TO THE FUTURE

In addition to gravitational waves, examples of Thorne's warped universe include the Big Bang, black holes, wormholes, and time travel. We've already seen that the Big Bang is writ large in the CMB. LIGO has already detected gravitational waves. We've also obtained images of supermassive black holes. Might this mean that wormholes and time travel are next in line for discovery? Maybe not, but it's also worth remembering that nearly every discovery that upended common knowledge was prefaced by or met with booming skepticism. Sometimes those who labor in the trenches of science are the loudest skeptics. Wormholes and time travel are indeed tricky to reconcile with the known laws of phys-

This computer simulation shows how the collision of two black holes 1.3 billion years ago—detected for the first time ever by the Laser Interferometer Gravitational-Wave Observatory (LIGO)—would appear to our eyes if we managed an up-close look.

ics, but that hasn't stopped great minds—whether physicists, philosophers, science fiction writers, or Hollywood directors— from pondering a universe compatible with these possibilities.

Just because an idea is crazy, though, doesn't make it true. The trash bins of science hypotheses overflow with wrong ideas. You'll never read about those, because they're all authentically crazy. So, feel free to adopt a healthy skepticism of nutty new ideas.

continued on page 258

BLACK HOLES IN THE MOVIES

Kip Thorne carved his name into pop culture when he worked with director Christopher Nolan on the sci-fi megahit *Interstellar*. The film showcases with stunning accuracy the strange and distorted phenomena permitted by relativity. Wormholes, black holes, extra dimensions, and time dilation are all plot points in the storytelling.

One of the more remarkable scenes depicts Gargantua, a massive black hole the protagonists encounter elsewhere in the galaxy. Thorne worked closely with the visual-effects team to produce what might be expected when facing a real black hole: a broad, radiant halo surrounding the central dark shadow with a thin band of light across that dark expanse— like an eerie Saturn. What we see is not the black hole but the surrounding light caught in its gravitational grasp. The ring across its midsection is whirling energy, and the rim of the shadow sphere is the distorted light behind the black hole, as seen from the front.

Interstellar was released in 2014, five years before astrophysicists would publish the first ever actual image of a black hole, lurking deep in the center of the giant galaxy M87. That supermassive specimen—more than six billion times the mass of the Sun—is located 53 million light-years away.

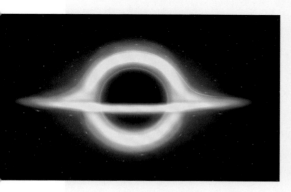

An illustration of the black hole depicted in the 2014 film *Interstellar*

You'll notice a few key differences between these two portrayals of a black hole. Clearly the resolution capacity of the Event Horizon Telescope cannot compete with the Academy Award–winning visual effects. Ignoring that, you'll notice that the crossbar ring is conspicuously absent on M87. That's not an oversight but the result of a different vantage point. Imagine viewing Saturn from top down or bottom up: The rings would appear as a circular halo with no crossbar in sight.

The first image of a black hole, captured using Event Horizon Telescope observations of the center of the M87 galaxy

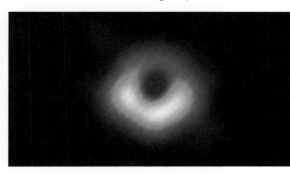

You'll also see that Gargantua is uniformly bright, whereas one side of M87's black hole is brighter than the other. That's the Doppler effect in action. One side appears brighter because the pool of photons is blueshifting, or moving toward the viewer, while those on the dimmer side are redshifting, or retreating as they swirl behind the black hole. Thorne, well aware that a black hole might appear this way to an onlooker, elected to sacrifice that bit of science for Gargantua.

This is an example of the "Tiffany problem" that crops up in movies. Coined by Welsh author Jo Walton, the phrase usually refers to historical works of fiction but can easily apply to sci-fi. "Tiffany" probably sounds like a modern American name, popular in the late 20th century but not much earlier, right? Wrong. Actually, the name originated in Europe in the 12th century. So, a story about someone named Tiffany, set in France during the Middle Ages, would be a perfectly legitimate choice. But that choice would raise too many eyebrows by viewers who think they know about the legitimacy of the name in time and place. It would seem too incongruous and effectively distract the audience from the main story.

Similarly, if *Interstellar*'s Gargantua appeared noticeably brighter on one side than the other, viewers would question the effect. Whether in science fiction or elsewhere, the authors of the best told stories know when and where to sacrifice truth if accuracy for the sake of accuracy might distract or confuse. As Mark Twain once said, "Get your facts first, and then you can distort them as much as you please."

H. G. Wells's cult classic
The Time Machine (1895)

continued from page 255

H. G. Wells popularized time travel as a science fiction plot device with his 1895 book *The Time Machine*. The protagonist of the story, a scientist and inventor, creates a device that transports him hundreds of thousands of years into the future. Although the story itself is remembered for its social and political commentary, it also incorporates Wells's provocative conception of time.

Wells poses a philosophical question: "Can a cube that does not last for any time at all, have a real existence?" the novel's time traveler asks his companions. "Clearly," he continues, "any real body must have extension in four directions: it must have Length, Breadth, Thickness, and—Duration . . . There are really four dimensions, three which we call the three planes of Space, and a fourth, Time."

Though Wells was scientifically literate and up to date on the current physics, this passage shows remarkable thinking, given that equations to support it did not yet exist—and wouldn't until the publication of Einstein's papers.

Einstein's 1905 special theory of relativity unlocked the first secret to time travel: the constant speed of light. Isaac Newton presumed a universal—or absolute—time, meaning that everything everywhere experiences and observes time the same way. Of course, Newton didn't have the faintest idea that other solar systems, let alone galaxies, awaited discovery. But if he had, he would have assumed that an extraterrestrial species in a galaxy far, far away could synchronize a clock with ours, and that, barring any technical problems with the clock itself, both clocks would remain in sync and we would all happily agree on the time and date of any event in the universe. However, Einstein's special theory of relativity reveals that simultaneity of any kind is a mere illusion, and that time itself varies from one observer to the next.

If an Intergalactic Space Olympics took place in Newton's universe, and if alien and human viewers across myriad planets could all witness the events from home, everyone would agree on the duration of the 200-meter dash. But in Einstein's universe, every zone of spacetime would clock a different winning time from that measured by the official timers at the Olympics. The sprint that might take 20 seconds from the point of view of the sprinters could carry on for years from the point of view of an alien world.

Spacetime imposes a speed limit to which everything must accommodate. But it does so in freaky, seemingly impossible, ways. Welcome to relativity, where time is—you guessed it—relative.

To understand why our world behaves this way, we can invoke a few simple thought experiments. Say a car with glaring headlights zooms toward you as you stand motionless. If a passenger in that car throws a ball to you out the window, the ball will arrive at the speed of the throw plus the speed of the car. What about the photons from the headlights? Shouldn't the headlights also move toward you at the speed of light plus the speed of the car? No, photons travel no faster than the speed of light, regardless of how fast the car itself may be moving.

> Spacetime imposes a speed limit to which everything must accommodate. But it does so in freaky, seemingly impossible, ways. Welcome to relativity, where time is—you guessed it—relative.

continued on page 262

DEATH BY TIME MACHINE

"Where are we?" asks one time traveler to another as they emerge from their futuristic time machine. "You mean *when* are we?" the companion predictably retorts. You'll often see that joke in time travel tales, films, and TV shows, even though the punch line exposes the writers' misunderstanding of the nature of spacetime. In fact, almost every time travel movie you've ever seen gets it wrong.

Here's a more accurate depiction of that classic scene: Two time travelers emerge from their futuristic time machine and promptly suffocate in the vacuum of space. THE END.

Unless you're traveling mere hours into the future, you'd better hope your time machine doubles as a spaceship. Chances are, when you arrive at your designated time and date, Earth will have long departed the coordinates you left as it continues along its orbit around the Sun at 67,000 miles an hour. Whatever time and date you choose, you must also specify a location to go along with it. You might attempt to solve this problem by traveling exclusively at whole-year intervals, to ensure that Earth can be found at the same point in its orbit around the Sun as when you left. The 1985 film *Back to the Future* snuck out of this problem by choosing to go back 30 full years rather than, say, 30 years and a week.

But hold on. Even if Earth was in the expected part of its orbit, you have to account for Earth's rotation, including the fact that it wobbles on its axis. At middle latitudes, Earth's surface rotates at about 800 miles an hour. You risk reemerging in the middle of the Pacific or in the sweltering Mojave Desert. So let's say you invoke that level of precision into your spacetime calculation as well. You're still not in the clear.

The Sun itself, and indeed the entire solar system, orbits the center of the Milky Way at more than 500,000 miles an hour, making one full loop every 230 million years. So you must account for that as well. Surely now you'll be able to land at the exact same spot on Earth where you entered your time machine.

Not so fast.

The Milky Way, too, moves through space. We and the Andromeda galaxy are falling toward each other at 250,000 miles an hour, across a distance of two and a half million light-years. So, unless and until you can factor in the movement of Everything over time—in other words, unless your time machine is a space–time machine—your time travel will be a time leap straight to your own death.

In this fan art, the TARDIS of the *Doctor Who* series jumps through space.

continued from page 259

Now imagine something faster and farther away from you, like an airplane. Passengers inside the airplane can jump up and down, walk around, or throw a ball from one end of the plane to the other, all at what they would consider regular speeds. Their reference frame is locked to the interior of the airplane. As long as the plane remains in constant motion, without changing direction, speeding up, or slowing down, the plane is their world, relatively speaking. On Earth, however, the ball they're tossing from the back of the plane to the front moves at the speed of the plane plus the speed of the throw. Meanwhile, the photons emitted by the plane's navigation lights move at light speed, never more, same as the headlights of the speeding car.

Now for the freaky part. In one of your science classes, you might have been told that

$$d = v \times t$$

d (distance) equals v (velocity) multiplied by t (time)

If the speed of light (v) is constant everywhere always, then t and d must change to compensate. In other words, for two different distances at the speed of light, time itself varies. This is our first and most fundamental equation for time travel. It's the only one you need to know as we delve deeper through spacetime and beyond. Herein lies the beauty and simplicity of mathematics as the language of the universe. Bizarre ideas such as wormholes or time travel or teleportation are sometimes described as "mathematically possible." The equation $d = v \times t$ is an example of what that means.

Here's another thought experiment. Imagine a spaceship zooming overhead at nearly the speed of light. Inside that space-ship, an astronaut stands in the middle and holds two photon guns, one in each outstretched arm, aimed at opposite ends of the spacecraft. If she pulls both triggers at the exact same moment, then—from her perspective—the front and back of the spaceship light up at the same instant.

From your perspective on Earth, however, that's not what happens. You see the back end of the spacecraft light up sooner than the front end. Who's correct? The mind-boggling answer is: both. How can this be?

At the same moment the photon guns release the light beams, the ship itself is carried forward. The back end moves rapidly toward the oncoming light beams from the photon gun aimed in its direction, while the front end rapidly retreats from the light beams aimed by the other gun. As a result, although the speed of light remains constant, the distance covered by the two light beams differs. The distance to the back of the ship becomes smaller than the distance to the front. From the perspective of an Earthling, the spaceship also appears smaller as it shrinks from back to front along the direction of its motion. At the front end, the distance becomes greater, so the light takes more time to reach it.

Now, say our astronaut clicks on the bathroom light to look in the mirror hanging along the side wall of the ship. From her point of view, the photons travel directly from the lightbulb to her face, to the mirror, and back to her eyes on the shortest possible path. But from the perspective of the Earthling, the light takes a longer, diagonal path toward the mirror as the spaceship hurtles forward and away from the light beams. The light, with its finite speed, must take a bit more time catch up.

As with the photon gun, the Earthling observes the photons of that bathroom light traveling for a longer period of time than the period the spacefarer observes. They are, yet again, both correct. From the perspective of the Earthling, everything that happens in that spaceship really happens more slowly. The

Radiation and high-energy particles from a distant star in
deep space interact with the atmosphere surrounding Earth,
causing a cascade of subatomic particles.

clock on the wall takes longer to tick off seconds. The astronaut's heartbeat slows. She lands back on Earth a teensy bit
younger than the Earthling, just for having moved faster through
space. There you have it: time travel into the future. In physics,
this phenomenon is called time dilation, and astronauts do it
all the time.

If the spaceship carries on past Earth at 95 percent light speed
for one year, the time delay accumulates. During the course of
that journey, the Earthling experiences three years for the astro-

naut's one. If the spaceship stops, turns around, and blasts back to Earth at the same speed, taking another year to do so, the earthbound observer will now be six years older, while the astronaut will have aged only two years. Both individual realities, however irreconcilable they may seem to our feeble intuitions, are accurate and true. This is not an illusion. It's physics.

But if the math can't convince you of this simple yet profound truth, perhaps the story of the muon will.

Every second of every day, showers of cosmic rays—charged subatomic particles careening through outer space—slam into Earth's atmosphere at 99.99 percent the speed of light. When they collide with our dense atmosphere, they rapidly bust loose into even tinier charged particles. The ensuing shrapnel consists of many unstable particles, including muons, which resemble electrons but carry 200 times their mass. We know from the behavior of muons in particle accelerators that, left to their own devices, they decay in about two-millionths of a second—about 150,000 times faster than you can blink. The average rate of decay is measurable, precise, and, most important, predictable. Despite their short life spans, muons move so fast that they can still travel nearly half a mile from birth to decay.

However, muons are birthed in particle showers far higher in the atmosphere than half a mile, around 9 or so miles above sea level. Consequently, they should decay long before they reach our instruments. And yet, despite their brief, precarious life spans, muons routinely reach Earth's surface. How, then, can they possibly cross every layer of Earth's atmosphere and collide with our detectors on terra firma if they're supposed to have been decimated many miles overhead? The answer is special relativity.

Recall that our spaceship appears to shrink from the perspective of the Earthling who's watching it fly past. But from

THE VIEW OF HYPERSPACE

We can now say that both the spacefarer and the muon traveled into the future relative to Earth time by traveling at nearly the speed of light. In fact, we all travel into the future, at exactly one second per second, or one month per month, or one year per year. We move into the future at the same rate relative to ourselves, to our own heartbeats, whether or not we move at the speed of light. That's why astronauts won't notice their clocks slowing down and Earthlings won't notice their own clocks speeding up. Both clocks are accurate. If we plot their worldlines side by side, we will see that the astronauts moved more through the space axis than through the time axis, as compared with those on Earth. When their worldlines reconnect, one will have lived longer on the time axis and be that much older.

Ten years after publishing his early work on special relativity, Einstein extended it to incorporate gravity as well, unifying space and time. This later work, dubbed general relativity, declared that just as time and distance warp at extreme speeds, so too do they warp in extreme gravity. Einstein's new equations meant there were now two ways you could travel into the future: move at higher speeds or experience extreme gravity. Either way, your time and distance distort relative to an observer, forcing your personal clock to tick more slowly.

General relativity also introduced a universe in which an object with enough mass could warp space so drastically that nothing would be able to escape. Black holes provided the perfect conceptual playground to ponder the fluidity of time, but until recently, hardly anyone thought that such an object could exist in nature.

Einstein himself was more skeptical about finding a black hole than about detecting gravitational waves. Nevertheless, a black hole was discovered decades before LIGO's groundbreaking discovery of gravitational waves—waves caused by the collision of two black holes.

the perspective of the astronaut inside the spaceship, it is Earth that shrinks. Yet again, both are correct. The muon survives multiple miles of atmosphere during its half-mile life span because, from the muon's perspective, special relativity shrinks Earth's atmosphere to less than half a mile. This allows the muon to reach ground safely before decaying, extending its otherwise brief life span. As with time, the length of any object—the distance it spans, which we think of as its size—is relative. Surely this must be an illusion. How can one thing be two different sizes?

To invoke the idea of illusion implies the existence of a single ultimate truth that has been fumbled by our senses. But the fallibility of our senses has no place in special relativity. Just

Time and distance both warp at extreme speeds and in extreme gravity.

because time and length are relative does not mean they're not real. In fact, relativity tells us just the opposite: Not only are they real; they are real in more ways than ever before imagined. From our perspective, the muon itself shrinks as the time it spends in transit, and thus its life span, lengthens.

All told, the detection of muons on Earth's surface is the best, naturally occurring, constant, and directly measurable evidence we have for time dilation. And it dilates by the exact amount predicted by relativity. There's simply no other explanation for why this little smidge of matter could ever survive a descent to Earth.

There's an even more precise way to test the theory of time dilation on our muon or any other particle. Once we know the particle's precise decay rate, we can send it through a particle accelerator such as the Large Hadron Collider. That particle will live longer inside the accelerator than it does at rest—once again, in exact accordance with the equations of relativity.

BLACK HOLES

Combine Newton's theory of gravity with his idea that light is composed of corpuscles. Now add the knowledge that the speed of light is finite. Stir it into the mind of a brilliant person, and you find reason to think that a star might be massive enough for its gravity to slow down those light particles so much that they cannot escape. In the late 18th century, an English astronomer, clergyman, and geologist named John Michell described this very idea. He called such objects dark stars. As prescient as his thought experiment was, his writings fell into obscurity for almost two centuries.

Einstein resurrected the possibility of what came to be called a black hole when he described a curved universe shaped by matter and energy. If we return to the analogy of the rubber sheet and weighted balls to imagine the fabric of space, we can see that the more massive the ball, the more the sheet warps in response, forming a gravitational well. This warping dictates all trajectories though space, including orbits.

Now envision a ball more massive than all the rest—one so massive that it creates a canyon in the sheet. Other balls that fall into this deep well cannot escape unless an enormous amount of energy comes to their rescue. Past a certain point in their fall, however, no amount of energy will save them from inevitable descent. That point is known as the event horizon, where gravity overwhelms the speed of light. At the precise cusp of the horizon, a photon attempting to exit will have been robbed of all its energy. The abyss that lurks within the event horizon is known only to fictional characters who have ever dared venture inside.

Until 1971, black holes were nothing more than an intriguing concept. The equations allowed for them, but no evidence for any such mysterious space warp had yet surfaced. It's a difficult job to detect something made only of gravity. An isolated black hole is, by definition, undetectable, although any matter unfortunate enough to have wandered too close will reveal its presence.

John Michell suspected you could detect one of his "dark stars" by observing another luminous star in orbit around it. Indeed, a blue supergiant star first tipped off astrophysicists to a black hole lurking some 6,000 light-years away among the stars of the Cygnus constellation. Astrophysicists caught sight of the star emitting some of the highest x-ray radiation ever seen from Earth. But the star wasn't the source.

> Until 1971, black holes were nothing more than an intriguing concept. The equations allowed for them, but no evidence for any such mysterious space warp had yet surfaced.

An artist's concept of a black hole drawing in and densely
compressing all surrounding matter

Something close to it—a mysterious companion—was. Decades
of observation would confirm, beyond a doubt, that the star's
strange counterpart was indeed a black hole, busily flaying the
orbiting blue supergiant star that had expanded a little too much
for its own good. The x-rays detected were the energetic death
throes of swirling stellar matter, shredding into its component
atoms as it superheated to millions of degrees, forming a luminous accretion disk—that glowing halo around the rim of a black
hole. The dying star unmasked a cosmic monster, otherwise
disguised in nothingness, as it gobbled down its atomic meal.

Many people tend to visualize black holes as a cosmic vacuum
cleaner. But no, contrary to popular belief, black holes don't

GPS AND ISS

f you rely on your smartphone apps to get directions, book a taxi service, order takeout, or find a date, you can thank Einstein.

The GPS satellites in medium Earth orbit must be calibrated to correct for gravitational time dilation. At their average orbital elevation of 12,500 miles, they are not as deep in Earth's gravitational well as surface dwellers on Earth. General relativity prescribes that their clocks tick slightly faster than ours. Local time and your coordinates on Earth are fundamentally conjoined, so if satellite operators did not correct for the slight time dilation accumulated by the difference in gravity, humans and Uber drivers would never find each other, deliveries would arrive at the wrong location, and it's anybody's guess who Tinder might pair you with.

But wait, what about the effect of special relativity? All satellites in medium Earth orbit must maintain a speed of 7,000 miles an hour to stay in free fall. So, shouldn't their clocks tick slower, not faster, than ours? Yes. But when you do the math (the relativistic calculations), the speeding of time at their location far from Earth's surface is greater than the slowing of time from traveling fast in orbit. General relativity, in that case, beats special relativity. As a matter of fact, it *is* rocket science.

What about the astronauts aboard the ISS, which lives in low Earth orbit? They move at five miles a second, completing a full orbit around Earth every 90 minutes. In LEO, a few hundred miles up, special relativity wins. So ISS astronauts age a bit more slowly than their friends on Earth. When NASA astronaut Scott Kelly landed back on Earth after his fourth and final sojourn, a marathon 340 days aboard the ISS with a Russian colleague, he had traveled a whopping five milliseconds into the future, compared with all of us on Earth, including his twin brother, Mark.

So there you have it: Humans will need to travel a whole lot faster for a whole lot longer to approach any meaningful time dilation consequences of special relativity.

suck. That blue supergiant wasn't being sucked into the black hole; its expanding outer layers were simply falling toward a gravity well, like the balls on the rubber sheet. The accretion disk surrounding a black hole is just the matter that has yet to fall in.

We now estimate that at least 100 million black holes wander the Milky Way galaxy— most puny, some colossal, their mass ranging from a few times that of the Sun to the giant at our galactic center, with four million times the mass of the Sun. Black holes of superhigh mass have earned the sensible name "supermassive black holes." Near a black hole's event horizon, one second could equal thousands or even millions of Earth years. In his 2014 companion book to the film he advised, *The Science of Interstellar,* Kip Thorne neatly summarizes gravitational time dilation with one sentence: "Everything likes to live where it will age the most slowly, and gravity pulls it there."

TIME TRAVEL: RETURNING TO THE PAST

"People like us, who believe in physics, know that the distinction between past, present, and future is only a stubbornly persistent illusion." —Albert Einstein

Time travel to the future is easy. As we've seen, the astronauts aboard the International Space Station do it all the time. Time travel to the past, however, requires bleeding-edge mathematics and near-impossible technologies.

And yet the equations are okay with it.

The faster you travel relative to an observer, the slower your clock will tick. At light speed, your clock will freeze. If you could travel faster than light, your clock would reverse. Think of it this way: If your spaceship sends out a beam of light while it's somehow traveling faster than light, you and your ship will beat that light beam. Loop behind it, and you could watch yourself race off into the emptiness.

It's fun to imagine what might happen if we traveled faster than light. Unfortunately, Einstein's equations prevent it. But does that make it impossible? If not, how then might we travel faster than light (FTL)? Scientists have concocted a few highly creative ways to beat a light beam without ever exceeding the universal speed limit.

FTL METHOD #1: THROUGH THE WORMHOLE

Physicists call it an Einstein-Rosen bridge, but you might know it as a wormhole.

If space can bend and warp, as Einstein described and as black holes and gravitational waves have proved, then perhaps the space between two distant points could fold in such a way that those two points could join up.

Think of it this way: If two ants on either side of a long piece of paper wanted to meet, you could watch them languidly crawl across the expanse toward each other. Or you could simply intervene in the situation and fold the paper in half for them, bringing them adjacent to each other. Likewise, if you could fold yourself over to a planet a hundred light-years away, you'd beat any beam of light by a whole century, assuming the beam didn't take the same shortcut. Through your wormhole, you could instantaneously inject messages or people into an otherwise unreachable worldline.

The 1997 film *Contact* (based on the 1985 novel by Carl Sagan) set a high bar for accurate physics portrayal in sci-fi movies. The story follows SETI (search for extraterrestrial intelligence) researcher Ellie Arroway's attempts to decode an

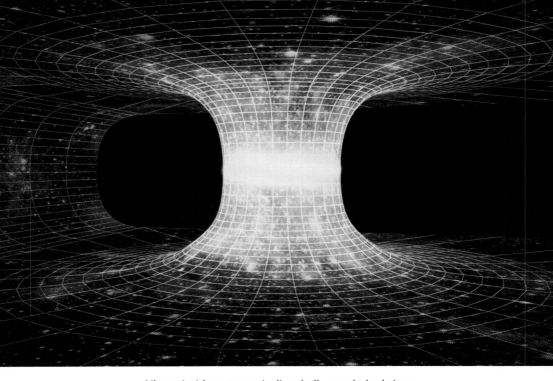

Like an inside-out cosmic disco ball, wormholes bring a smorgasbord of possibilities to the interstellar party.

alien message, eventually leading to her journey through a wormhole to meet with the advanced extraterrestrial species on their faraway planet. Sagan consulted Kip Thorne, already one of the leading authorities in relativity physics at that time, on the best method to portal Ellie from one distant galaxy to another in the blink of an eye. Sparked by the desire to tell a thrilling yet accurate story, Sagan and Thorne offered the world wild new ideas never before thought to be compatible with an Einsteinian universe.

Long before Sagan turned to Thorne, other theoreticians had pondered wormholes. While assisting Albert Einstein at Princeton's Institute for Advanced Study, an American-Israeli physicist, Nathan Rosen, helped predict the possibility. Together they published a 1935 paper on the idea, which inspired the name Einstein-Rosen bridges. Two decades later, another American physicist, John Wheeler, coined the term "wormhole" in a famous paper that would also prove them to be a paradox. Anything attempting to traverse the point between the two connected openings, he realized, would trigger its immediate collapse. A wormhole is thus fundamentally (and lethally) unstable.

Extraordinary questions and conundrums arise at the frontiers of cosmology. One such might be, "What type of exotic matter might an infinitely advanced civilization plug a wormhole with?"

Thorne knew from both Wheeler's and his own research that a wormhole seeks to close in on itself so rapidly that not a single photon can pass through—at least, not intact. To keep it open long enough for Dr. Arroway to cross safely, thus avoiding an abrupt, undesirable end to the story, the advanced civilization must thwart the portal's urge to collapse. They would need to deploy some type of matter imbued with repulsive, negative energy. Kip Thorne called it exotic matter, as nobody had yet observed anything with such properties.

Notice that Thorne did not name it impossible matter or imaginary matter. No. In all fields, the best theoreticians, philosophers, and thinkers know to keep a door ajar for the unprecedented and improbable. A decade later, cosmology would confirm the existence of dark energy, an invisible substance with negative gravity: precisely the repulsive properties John Wheeler and Kip Thorne had invoked.

Quantum physics offers a variety of intriguing candidates for the exotic matter required to prop open a wormhole. Let's just assume that advanced aliens harnessed whatever that stuff is and used it to set one up in our backyard. (This, by the way, is a premise of *Interstellar*. Using a fairly far-fetched method, Thorne postulated how to build a wormhole time travel machine.)

Imagine a wormhole as a Slinky connecting two points. If that Slinky were stupendously long and stretchy, one end—one

A COSMOLOGICAL CONSTANT

Through the end of the 20th century, Einstein's cosmological constant continued to raise questions. Salient among them: Even if the universe is expanding, as we clearly observe, might gravity pull everything back together, as Einstein feared?

In 1998, two independent teams of astrophysicists observing distant supernovae through the Hubble Space Telescope realized that the explosions appeared far fainter than they should if cosmic expansion were indeed slowing. In fact, their analysis proved the opposite: that the expansion is in fact accelerating, a finding that earned them the 2011 Nobel Prize in Physics. The only explanation, and still the best explanation to date, is that a mysterious antigravity entity, constituting about 68 percent of the universe, overrides the inward gravitational pull of all matter.

Today we refer to that entity as dark energy. We have no idea what it is or where it came from, but we know it exists, busily shaping the fabric of space and time. Instead of an Einsteinian doorstop gently propping the universe open and preventing collapse, dark energy is more like a strong gust of wind blasting it open wider and wider at a terrifying pace. In any case, Einstein's dreaded cosmological constant turned out to be real, which means it was a huge blunder to have called his introduction of the cosmological constant his biggest blunder. In other words, even when Einstein was wrong, he was right.

The long-term consequence of this expansion is the thinning of spacetime until even the cosmic microwave background dilutes to nearly nothing. Once the atoms that constitute the cosmos no longer collide, the universe will be reduced to cold, dark silence.

"mouth"—might dilate time relative to the other. Let's say an astronaut friend asks Earth-dwelling you to guard one mouth of the ultrastretchy Slinky wormhole while she carries the other mouth on her journey through space at nearly light speed. If she returns one year later, as she experienced that one year, she will encounter a 10-year-older you and a 10-year-older wormhole mouth. Your friend and her end of the wormhole, however, aged only a single year.

Now, if you step through her end, you will encounter your younger self waiting patiently for your friend's return. Your younger self, witnessing your 10-year-older self suddenly emerging, can step through the same wormhole mouth to arrive 10 years into the future. If you keep that wormhole portal open, then any future generation of humanity could now step through it into the moment the wormhole was created. If we carried on the tradition of forming a new wormhole every year, future humans would have access to a kind of spacetime elevator that opened up into whatever year they'd like—provided that exotic matter, harnessed for this very purpose, kept all the wormholes pried open.

Given all these constraints, are wormholes possible? Today's leading figure in wormhole science, Kip Thorne, thinks the answer is, Probably not. But in a 2019 lecture at Cardiff University, he told the audience, "When speculating beyond the frontiers of firm knowledge, I've been proven wrong many times—sometimes spectacularly. So don't take my pronouncements too seriously." Ever the theoretician, Thorne once again leaves the door of cosmic possibility slightly ajar.

> Are wormholes possible? Today's leading figure in wormhole science, Kip Thorne, thinks the answer is, Probably not . . . Ever the theoretician, Thorne leaves the door of cosmic possibility slightly ajar.

FTL METHOD #2: FIRE UP THE WARP DRIVE

The original *Star Trek* series of the late 1960s popularized the science fiction device called the warp drive, which enabled starships to explore, befriend, and attack extraterrestrials across the universe. In the 1980s, *Star Wars* introduced the hyperdrive, which similarly enabled faster-than-light travel. These fun plot devices employed fanciful terms and fictitious fuels to describe the fantastical technology.

Warp speed stayed strictly planted in the scripts of sci-fi authors until Mexican theoretical physicist Miguel Alcubierre penned the 1994 paper "The Warp Drive: Hyper-Fast Travel Within General Relativity." In it, he proposed a method that was perfectly, though puzzlingly, permissible under special and general relativity. Like Kip Thorne, Alcubierre called upon exotic matter—in this case, to expand space behind a spaceship while contracting space in front of it.

We know that relativity prohibits matter from traveling faster than light within the fabric of spacetime. But it does not prohibit spacetime itself from stretching at whatever speed it so pleases. An Alcubierre drive would create a bubble of local spacetime around a spaceship that could travel at any speed through surrounding space. Just as a galaxy is carried along for the ride as the space around it expands, the spaceship and its crew needn't move at all as the encapsulating bubble carries the vessel along like a surfer on a wave. With enough exotic matter, powered by a warp-drive engine, space could scrunch up in front of, and stretch out behind, any spaceship traversing space at otherwise impossible speeds without violating any laws of physics.

Just because something could work, or is consistent with known physics, doesn't make it a realistic possibility. Alcubi-

An Alcubierre drive could theoretically contract the space in front of a spacecraft and expand the space behind it, permitting the craft to travel faster than the speed of light without violating any laws of physics.

erre's original proposal required more energy than could be acquired from all the mass in the observable universe. More recent studies have knocked the requirements down to more plausible, though still improbable, amounts of negative energy. And that energy is yet to be observed, let alone harnessed.

FTL METHOD #3: DEPLOY THE TACHYONS

Physicist Gerald Feinberg introduced the term "tachyon," from the Greek *tachys* (fast), in his 1967 paper "Possibility of

continued on page 282

VIEWS FROM HYPERSPACE

A *Star Wars* film would be incomplete without the iconic view of entering hyperspace: "Go strap yourselves in, I'm gonna make the jump to light speed!" Han Solo warns his comrades in the *Millennium Falcon* before flipping on the hyperdrive. In a flash, every star in the *Falcon*'s forward field of view streaks toward the camera in ribbons of white and blue. It makes for a stunning, quintessentially *Star Wars* spectacle of FTL travel. An accurate warp drive display might appear equally stunning but very different.

As your ship's scientifically accurate warp drive approaches light speed, the wavelengths of the stars' radiation contract so drastically (because of the Doppler effect) that the starlight shifts to blue, then violet, then to invisible ultraviolet, and thence into yet other invisible and increasingly harmful frequencies. If your starship was so poorly conceived that it included an enormous glass viewport, you'd want to keep it closed to protect yourself and your crew from the oncoming x-rays and gamma rays.

But if you did dare a peek out the forward window, space would not appear dark—for, as we know by now, the cosmic microwave background permeates all of spacetime. The long, cold wavelengths that are invisible to our unaided eyes would shift as well—shortening so considerably that they would approach the visible spectrum. The entire sky would then glow blue, awash with the Big Bang's first light.

Star Wars: Episode IV—A New Hope (1977)

continued from page 279

Faster-Than-Light Particles." Feinberg found a loophole in Einstein's equations that would enable a particle to travel at the speed of light, so long as it did so always and forever. He dubbed the particle a tachyon as the counterpart to our regular, everyday, slower-than-light particles, sheepishly dubbed "tardyons."

Strictly speaking, special relativity inhibits a particle from accelerating past the speed of light. Feinberg proposed that the laws needn't apply to any particle that was both birthed at FTL speed and carries on forever at FTL. Yes, tachyons must also obey a speed limit—a slow speed limit. As long as it never traveled slower than light—crossing that speed barrier from above to below—then no equations could prevent it from traveling faster.

Cause and effect slip away in a world where tachyons are permissible. A tachyonic messaging app would deliver texts

This hypothetical particle, the tachyon,
moves faster than the speed of light and travels backward.

before a sender sends them. Imagine this text message popping up on your tachyonic phone: "Watch out for that banana peel!" When you look down, sure enough, a banana peel is squashed between your boot and the floor. The message comes from your friend down the hall, who only moments earlier witnessed you slip and fall.

BREAKING AND REPAIRING CAUSALITY

To send a message, a person, or a single photon anywhere faster than the speed of light will require the help of an incomprehensibly advanced civilization, unimaginably vast stores of exotic matter, or both. But these technological roadblocks are mere hiccups in the face of the truly insurmountable challenge of causality.

With a faster-than-light message device, the following scenario could occur: You captain a spaceship traveling away from Earth at 99 percent the speed of light when you receive a tachyon text message from an FTL transmitter based on Pluto. The message reads, "The Death Star just obliterated Earth."

On your worldline, however, Earth has not yet been hit. You could turn on your wormhole-maker, punch in the coordinates to the Death Star, and save humanity in the nick of time.

But wait, if you save humanity, then who knew to send that transmission? Furthermore, you received that transmission before anyone even sent it. The FTL message broke the principle of cause and effect.

That's what we call a paradox. In physics, paradoxes don't and can't happen, period. Let's plot some worldlines to understand why that message arrives before Earth explodes.

The time axis (y) on our two-dimensional worldline graph is vertical, and the left–right space axis (x) is horizontal. A three-dimensional worldline graph would have a third space axis (z)

extending through the back of this book and forward through your body—but as that axis functions no differently from our x space axis, we'll ignore that one for now. If we remain stationary, we don't move along the space axis in either direction and simply travel along the time axis upward in a straight vertical line (toward the future) at a rate of exactly one second per second. We can plot that for an unmoving Earth, and an unmoving Pluto (relative to one another).

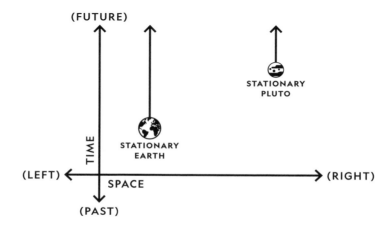

How about the opposite scenario? If an object moves along the space axis in a perfectly horizontal line, and not along the time axis at all, that means that the object is teleporting—via a wormhole, perhaps.

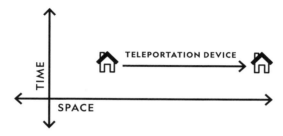

All objects in the universe that we know of, save in science fiction movies or hypothetical thought experiments, can only

QUANTUM WORMHOLE SOUP

Most physicists think we're unlikely to stumble upon a naturally occurring wormhole. Unlike black holes, billions of which loiter in our galaxy, wormholes can't exist on their own—at least, not large ones.

Recall that a virtual particle can pop into and out of existence even in the darkest, emptiest corner of space. These spontaneous, unpredictable energy fluctuations are the primary reason we can't say with certainty that anywhere is truly empty. John Wheeler, the wordsmith-physicist who coined the term "wormhole," also coined the name for the spontaneous perturbations that may underpin and permeate space: quantum foam.

Wheeler proposed that a roiling soup of virtual particles, wormholes, and other distortions of spacetime flourishes at Planck scale beneath and within the seemingly simple, predictable macro universe we inhabit. The Planck scale is the tiniest measurement we know. It is so small that descriptive words and analogies fail. To begin to grasp the degree of its minuteness, we might borrow the American physicist Brian Greene's analogy: If a single atom were blown up to the size of the entire observable universe, then the Planck scale would be the size of an average tree on Earth. "So great would be the fluctuations," Wheeler proposed in his memoir, *Geons, Black Holes, and Quantum Foam,* "that there would literally be no left and right, no before and no after. Ordinary ideas of length would disappear. Ordinary ideas of time would evaporate."

---

move through the left and right space axis at the speed of light or slower. Instead of a horizontal line (teleportation), the worldline of an object is drawn at a 45-degree angle above the time axis in all directions. This is our light cone, and anything in our future or present rests within this cone of

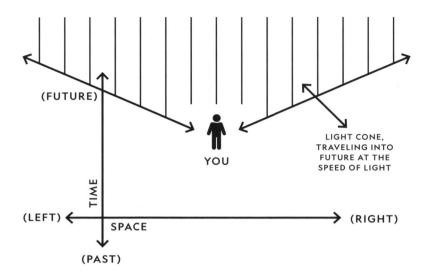

(FUTURE)

LIGHT CONE, TRAVELING INTO FUTURE AT THE SPEED OF LIGHT

YOU

TIME

(LEFT)

SPACE

(RIGHT)

(PAST)

possibility. We can also draw a 45-degree angle below the object to represent all events that could have been perceived in its past.

Let's now return to our tachyon scenario. Here's where it gets tricky. That FTL transmission not only teleports from one point to another on the space axis but also transcends the confines of the light cone. In other words, it can arrive within another light cone where the message itself had not yet been sent. When Earth explodes, radiation from that event traveling at the speed of light eventually enters Pluto's light cone. Then, sources on Pluto shoot a tachyon text message to your spaceship moving relative to both Earth and Pluto. On your worldline, you receive that text message long before the light of the explosion reaches your ship. But even stranger than that, you receive the message before Pluto ever sent it. In fact, you receive the message before the explosion itself ever happened at all. Your spaceship's worldline relative to Earth, Pluto, and the explosion is not the same. Your light cones of possibility are different. So, we ask again, who sent that tachyon text message if it arrived before it was thought up and typed out?

This is the insurmountable paradox of causality, also known as the "grandfather paradox." If you could somehow travel to the past and prevent your grandparents from meeting, then they could never give birth to your mother, who could never give birth to you. And if you weren't born, then you couldn't time-travel into the past in the first place. If everything that ever happened has already happened, then we cannot change the past.

Stephen Hawking wittily proposed the need for a chronology protection agency to keep the universe safe for historians. Indeed, plenty of time travel tales play on the concept of policing time travelers to prevent paradoxes. Within the Marvel comics and Marvel Cinematic Universe, the Time Variance Authority oversees the Sacred Timeline. The Netflix series *Umbrella Academy* has the Temps Commission, which serves as their time-preservation FBI. And of course, the Time Lords of the *Doctor Who* TV franchise are tasked with all things "wibbly wobbly timey wimey" within their universe.

Cause and effect appear to rule the universe, reigning supreme over the foamy quantum bubble bath imbuing all of spacetime. If anything is capable of breaking cause and effect, then everything we understand about our universe breaks along with it. There's no simple way around the causality paradox that arises with FTL travel and time travel into the past.

However, there's one worldline weirdism that appears perfectly consistent with all known laws of physics yet evades the tricky problem of causality: the causal loop, or bootstrap paradox. A causal loop does not require a chronology protection agency to preserve our timeline from being undercut by backward time travel, because the causal-loop timeline requires our return to the past to shape the events.

Recall the example of the banana peel and the tachyon text message. You're walking down the hall when you receive the message "Watch out for that banana peel!" Startled, you stop in

PARTY OF ONE

O n June 28, 2009, Stephen Hawking, one of the most famous and beloved scientists in modern history, threw an extravagant party. He announced the event on television and across the internet, inviting every person in the world to join him. He sent the invitations far and wide, specifying the precise date, time, and GPS coordinates for the grand affair. And yet nobody showed up.

How could such a bash turn into such a flop? Hawking didn't send out the invitation until after the event—on purpose. It was a party for time travelers, complete with balloons, champagne, and a welcome sign.

Hawking's party of one was an answer to his own famous conundrum: If we can time-travel to the past, where are all the time travelers? If anybody had ever invented a time machine that could take them to a specific time and place in the past, surely somebody would have arrived at that party to sip champagne with the great Stephen Hawking.

If you're prone to conspiracy thinking, you might wonder whether time travelers keep all their journeys a secret. Maybe they avoid seeing anyone or touching anything in the past. Maybe there's a Temps Commission that forbade attendance at Hawking's event, or the Time Variance Authority pruned any variants present. Maybe a time traveler was hiding behind the curtains the whole time.

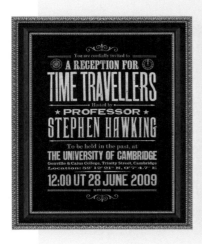

Unlikely. One thing human history teaches us again and again is that we're lousy at keeping secrets. Scientists might be the lousiest of all. As Benjamin Franklin once wrote in *Poor Richard's Almanack*, "Three may keep a secret, if two of them are dead."

The iconic time machine in Simon Wells's 2002 *The Time Machine*

your tracks. If you'd kept striding normally, your left foot would have overstepped the slippery peel. But now that foot lands squarely atop it. You slip and fall. The friend who sent the message in an attempt to save you from your fate made your stumble inevitable—a self-fulfilling prophecy. In a causal loop, you might attempt to change your timeline, but you cannot succeed, no matter what you do. The future is determined, and you shaped it yourself.

The 2002 movie version of *The Time Machine,* partly based on H. G. Wells's novel and directed by Wells's great-grandson, invokes the bootstrap paradox to heart-wrenching effect. After the brutal death of his fiancée, the protagonist invents a time machine to go back and save her life. He inevitably, devastatingly, fails. He returns to the past to force a new outcome—but

continued on page 292

HITLER'S MURDER PARADOX

The 1984 film *The Terminator* starred Arnold Schwarzenegger as a time-leaping, artificially intelligent, homicidal cyborg with a single mission: to kill the unsuspecting heroine, Sarah Connor. The Terminator (opposite) leaps back from a postapocalyptic future, dominated by an evil AI organization named Skynet that seeks to annihilate every human. Faced with a dedicated army of human resisters, led by a person named John Connor, the AI overlords try to prevent his birth by killing his mother, Sarah, before he's born.

Engaging the grandfather paradox will free Skynet to exterminate humanity. Gory brawls and brutal deaths thicken the plot as the Terminator pursues his mission (opposite). But if the AIs had any understanding of human physiology, they would know that plenty of time, energy, and bullets could have been saved by simply delaying the moment of conception. If Sarah Connor had conceived her child just a few hours later or earlier, a different sperm would almost certainly have fertilized the egg, and the John who led humans into battle would never have existed. Someone else might, but not John.

If it were possible to save the world from unspeakable atrocities by traveling back in time, many of us would elect to do so to kill Adolf Hitler and prevent the horrors he unleashed. The trope recurs so often as a plot device that it's become a topic of public debate: "Hitler's Time Travel Exemption Act" or "Hitler's Murder Paradox."

But if we simply kill Hitler, how can we be sure the events that follow the assassination won't be worse than what already happened? Operation Foxley, an assassination plan meticulously developed by Britain's Special Operations Executive, met opposition by some SOE personnel for that very reason. By 1944, when one or another of the Foxley options was to be carried out, Hitler's wartime blunders seemed preferable to the strategies of his shrewder generals who would presumably replace him.

In 1998, half a century after the war's end, the British government released formerly secret documents concerning Operation Foxley. With those detailed plans in hand, a time traveler could easily go back and carry out the mission. Anybody reading these words, however, knows that nobody has yet done so.

But there might be another, less violent way to ensure Hitler never takes power. By 1908, the young Hitler had twice been rejected by the Academy of Fine Arts in Vienna, his work having been deemed unsatisfactory. A time traveler might be wiser to go back in time and change that rejection letter into an acceptance letter. Perhaps the artistically successful Hitler would never have found time for the political ambition that would plunge the world into darkness and war. Perhaps the artist would have reveled in creation rather than destruction. Or maybe all he really needed was a hug.

continued from page 289

again, his fiancée is killed. Defeated, he realizes she cannot be saved, and that if she had been, he would never have invented the time machine that would try to save her. Their future was cast by the events that preceded it.

Another loophole out of the causality problem—similar to the bootstrap paradox—is offered by the jinnee particle. Proposed by Russian physicist Igor Novikov and his colleague A. Lossev, a jinnee's worldline, or existence, is a closed loop. It has no beginning and no end in spacetime. Named for the powerful beings that appear and disappear and metamorphose by magic in Islamic myth (from whose name derives the more familiar word "genie"), a jinnee particle could be an object, a person, or even a piece of information.

Picture yourself traveling back to the cobbled streets of 1804 Vienna. As you wander, you hum your favorite music—Beethoven's Fifth Symphony. Unbeknownst to you, the great composer, though plagued by increasing deafness, is walking nearby and overhears your humming. Moved by the powerful notes, he goes home and pens one of the most famous symphonic themes in history. If he hadn't written the notes, you could not have hummed them. If you had not hummed them, he would not have written them. In this example, the Fifth Symphony is a jinnee, with no clear origin. It always was and always will be, trapped in the cycle of its own timeline.

MANY, MANY, MANY WORLDS

The philosophical concept known as Occam's razor tells us that the simplest solution is nearly always the correct solution. The simplest solution to our causality problem is that time travel is simply impossible—that the laws of physics prohibit it. The next simplest solution is that backward time travel is indeed possible,

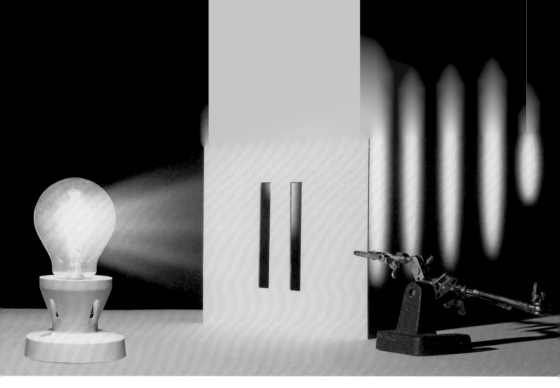

In the famous double-slit experiment, a light source shines toward a plate containing two narrow slits and creates an image on a screen behind the plate, demonstrating that photons behave as both a wave and a particle.

but only under the supervision of a chronology protection agency that keeps every stone unturned and prevents any alterations to the prewritten timeline. But do not despair, time travel enthusiasts. Cosmologists are nothing if not tenacious, especially when it comes to loopholes in Einstein's equations that might make a good story.

Another way out of causality calls on the quantum foam: the bubbling broth of quantum fluctuations of particles and perhaps even wormholes that flourishes at Planck scale. In the quantum world, where Heisenberg's uncertainty principle rules, everything is both a wave and particle. Wavicles exist in a state of probability; all options are always on the table. But the moment you measure it, the moment you identify where it is, the wavicle freezes into a particle at a single location, constituting a single possibility. What physicists now call the many-worlds interpretation (MWI) asks one question: What if all those quantum possibilities are as real as the single one that was measured? If all options are available and real, this multiplicity would manifest as many universes.

We know now that a photon is both a particle, as Isaac Newton hypothesized, and a wave, as Christiaan Huygens hypothesized. They were both right. At macro scales, we can't see this wave/particle duality, this wavicleness, but at subatomic scales we can test and observe it. Thomas Young's famous double-slit experiment in physics, which dates back to 1801, shows just how weirdly the physical world behaves when we observe it at the smallest possible scale—and even more so when we do not observe it at all.

> We know now that a photon is both a particle, as Isaac Newton hypothesized, and a wave, as Christiaan Huygens hypothesized. They were both right.

Let's say you have a special gun that will fire individual photons, one at time, at a barrier notched with two exceedingly narrow slits. The slits are small enough to allow each photon to enter one slit or neither, but not both. Each time you shoot the gun, the photon will, sure enough, choose one slit or neither, just as you would expect.

Continue the same experiment multiple times, and eventually your data will firmly point to each slit welcoming half the visiting photons. If each photon left behind an imprint indicating where they landed, you could look behind the slitted barrier, to find two neat imprint lines, one behind each slit. If you cover one slit, you will observe a single neat line of photons. This is all perfectly normal. It bespeaks the rational behavior of any particle. If you fired small paintballs instead of photons at a similar type of barrier, you would see the same results: two neat lines behind the notched barrier where each paintball landed.

But here's where it gets inexplicably strange. Let's now say you head for a nearby taco stand for your lunch break and let the experiment carry on autonomously. You set the gun to continuously fire one photon at a time at the barrier. When you return, you expect to see the same pattern of two neat lines on the wall behind the barrier, just as before. Instead what you see is a cosmic conundrum yet unsolved.

Where you once saw two neat lines, you now see many. The line pattern decorating the back wall is the same pattern that you might see on the surface of a lake where two ripples meet. The pattern implies that two photons, moving in a wavelike pattern, collided with one another over and over again while you were out wolfing down tacos. Did your photon gun malfunction and start shooting out two photons at once?

No. What happened was that while you weren't looking, the single photon chose both slits at once and interfered with *itself*. Instead of behaving like a particle following a predictable pattern, the photon behaved like a wave. But that's not all. Not only did the photon choose both slits at once; it chose all options at

The Sacred Timeline as seen in Marvel Studios' *Loki* (season 1, episode 6)

once. It went through the left slit, through the right slit, and through neither slit.

This is not a thought experiment. Countless double-slit experiments have been carried out by meticulous scientists during the past century. Every single time, the results show that the very act of observation causes the photons to behave like particles and make a decision between the slits. When we look away or vacate the premises, they melt back into waves that opt for all options. Once wavicles are observed or measured, they exist as particles with a single location in spacetime—observation forces a photon into only one option or reality. Yet another example of a spooky quantum phenomenon.

American physicist Hugh Everett III, while a Ph.D. student of John Wheeler's, wrote a baffling thesis offering a new interpretation of the results demonstrated by the double slit. His idea first met with outright ridicule and dismissal and was not seriously considered until the final years of his life. Everett proposed that all the unmeasured wavicle probabilities are as real as the measured outcomes. At every instant of every quantum particle's decision, the universe branches off into a separate, parallel universe in which the unchosen option was chosen. He audaciously suggested that the simplest explanation for a particle's tendency to choose all options at once—its superposition—is that it literally chooses all options at once at all times, whether or not we measure it.

The observer sees the photon choose one slit and a copy of that observer sees the photon choose the other, but neither observer occupies the same universe, and neither is the same person. The implications of Everett's "many-worlds" explanation, as it is now called, set forth a worldview more humbling and more astonishing than heliocentricity, the expanding universe, and even the possible discovery of intelligent extraterrestrials. What Everett proposed is that ours is not the only universe,

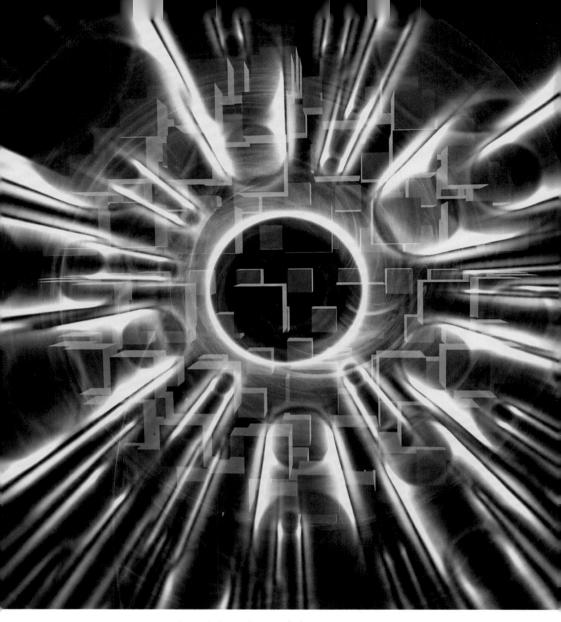

An artist's rendering of photon emissions

but merely one of an infinite number of universes. Our planet is merely one of infinite Earths, and you are one of infinite versions of the experimenter.

Every time a quantum physicist observes a photon entering the right-hand slit, at least two more universes branch off from their own: one in which it enters the left-hand slit, and one in which it chooses both slits. If every particle in the universe behaves this way, then, according to the many-worlds hypothesis, an incomprehensibly large number of worlds appears with every passing micro-moment. Everything that could possibly happen, happens. But we only know, and can only ever know, one outcome.

This idea should not be confused with the multiverse theory, which predicts, among other ideas, that our entire universe is a single bubble amid infinite other universe bubbles, all coexisting in spacetime; we could dream of one day opening a wormhole to one of these multiverses. A many-worlds interpretation of the universe, by contrast, says that we could never interact with any of the parallel worlds. It's not a question of their existing beyond the reach of our telescopes. It's that they exist beyond the reach of everything that constitutes our cosmos.

And so, if the many-worlds interpretation holds true, then a backward time traveler cannot violate causality because another world, where causality is preserved, automatically branches off. The many-worlds solution may not be simple, but it is the simplest explanation for the oddities at quantum scales.

WHAT ABOUT FREE WILL?

Ponder this: You loop back in time at FTL speed, whether by wormhole or warp drive, to the moment before you departed. To test the hypothesis of the many worlds, you decide to fire your plasma gun and blow up your own ship before it leaves. If the hypothesis holds true, then you will outsmart the causal paradox. The moment you travel back in time, your universe branched off into a different universe entirely. You jump back in time and explode a spaceship, but it was never your spaceship. You are writing a new timeline.

Within the terms of this hypothesis, a backward time traveler is also a parallel universe traveler. Every act that would otherwise alter an already established timeline in one universe would instead happen in a different universe, preserving the former timeline. The former "you" that would have gone back in time disappears from the timeline that would otherwise have been violated.

But what about the past "you" that the future "you" left behind? Wouldn't that former you still choose to zoom off at light speed to go back in time to explode your own ship, again causing a universe split? In other words, even if the original universe undergoes a split at the moment of causality violation, nothing changed in that universe to prevent the original outcome. Every neuron fired, every thought and memory and feeling that culminated in your decision, would still culminate in that same decision.

Was that decision predetermined by all the events that preceded it, including your genetic makeup, your childhood traumas, and your very consciousness?

Now we're staring at a central question: free will. It unites and recruits all sciences and philosophies. Is our future already written by the events of our past? Was a rigid and inevitable series of events set in motion by the Big Bang, or can we alter our timeline?

Even the most subtle of shifts by a time traveler on their own timeline could yield an entirely new reality. Statisticians call this phenomenon the butterfly effect, and scientists who work on chaos theory know never to underestimate the magnitude of cascading events that unfold from a single, seemingly innocuous event.

Imagine yourself looping back many years before embarking on the space adventure where you intend to test the many-worlds hypothesis. Like a small angle that opens wider and wider over longer and longer distances, any small change in

"Where we're going, we don't need roads."
—Dr. Emmett Brown, *Back to the Future* (1985)

your long-ago past could wildly alter your future. Maybe an adorable stray puppy appears in your past self's path, and so instead of heading to the bookstore, as you were planning to do that morning, you change course and take the puppy home.

Had you not encountered the puppy, you would have kept walking to the bookstore, where you would meet a new best friend who would spark your interest in poetry. And so, because you never met, you never went to the slam poetry contest at which another pivotal moment happened.

On and on the timeline changes, to such an extent that you never venture into space and so you never travel back in time to encounter the puppy in your path. You are a different person entirely, imbued with different memories, assailed by different fears, ablaze with different hopes, all imparted to your consciousness through differently wired neurons in your brain.

THE JOURNEY CONTINUES

What a journey it has been, this unending human quest to gain access to the heavens.

We have traveled far from our earthbound location and our egocentric notions of our place in the cosmos. We have come to understand the nature of the atmosphere—that unique blanket of gases that cushions and protects us and allows life to flourish on this planet. We have glimpsed our Sun, our Moon, and our neighboring planets, first by standing on Earth, telescopes in hand, and then by sending spacecraft up and out of our little shell, with each generation uncovering more and more mysteries farther and farther away.

We have evolved from counting planets in the solar system to counting universes in the metaverse, from questioning whether the Moon and Sun orbit our world to questioning causality in the spacetime continuum.

How much more can we boggle the mind? How much more is there to observe, conceptualize, and understand? What more will the cosmos teach us? On the bleeding edge of curiosity, where discovery meets mystery, we encounter a never ending parade of conundrums that we never could have imagined decades before—the delightful outcome of explorations into boundless space and endless time.

Scientific thinking always leaves the door ajar for the seemingly impossible. So perhaps we exaggerate—but only just a little—when we declare that infinity is only a moment's pause on the way to unlimited destinations that await us. For all we know, our cosmic journey has only just begun.

The James Webb Space Telescope's Near Infrared Camera captured this stunning mosaic image of the Tarantula Nebula star-forming region, its size estimated to be some 340 light-years across.

ACKNOWLEDGMENTS

We thank Avis Lang for her tireless and heroic edits of our early manuscript, once again, ensuring we say what we mean and mean what we say.

We also thank executive editor Hilary Black and senior editor Susan Tyler Hitchcock of National Geographic, whose creative intuitions and editorial wisdom served as guideposts along our own cosmic odyssey that became this book.

The rest of the publishing and design team at National Geographic did what they do best—turn pages of words into feasts for the eyes. They include editorial project manager Ashley Leath, creative director Elisa Gibson, designer Nicole Roberts, photo director Adrian Coakley, photo editor Katie Dance, and production editor Michael O'Connor.

Finally, we thank our friend and colleague Janna Levin for lending us her knowledge and counsel as a cosmologist, as well as a brilliant writer and communicator.

Lindsey Nyx Walker also recognizes Adrian Solgaard, for his unwavering support through the frets and joys that writers endure; Helen Matsos, for her guidance, friendship, and sage mentorship; Professor Ralph Engelman, Professor Curtis Stephen, and Professor Donald Bird, who sparked and encouraged her endless pursuit of finding and propagating the truth. And last but not least, Sue Ann Walker and Wallace Walker, for everything they are, and for instilling the love for words and exploration.

FURTHER READING

A few portions of this book are resurrected and heavily revised from three of Neil's essays published in *Natural History* magazine:

Tyson, Neil deGrasse. "The Coriolis Force." *Natural History,* March 1995.

———. "Tides and Time." *Natural History,* November 1995.

———. "Shocking Truths: If You Break the Sound Barrier, You Can Make Quite a Stir." *Natural History,* September 2006.

PART ONE

Galilei, Galileo. *Sidereus nuncius.* 1610.

———. *Dialogus de systemate mundi.* 1641.

Glaisher, James. *Travels in the Air.* R. Bentley, 1871.

Newton, Isaac. *Philosophiae naturalis principia mathematica.* 1687.

Tyson, Neil deGrasse, and Avis Lang. *Accessory to War: The Unspoken Alliance Between Astrophysics and the Military.* W. W. Norton & Co., 2018.

Weir, Andy, *The Martian.* Random House Publishing Group, 2016.

PART TWO

Gates, Jr., S. James, and Cathie Pelletier. *Proving Einstein Right: The Daring Expeditions That Changed How We Look at the Universe.* PublicAffairs, illus. ed., 2019.

Hamacher, Duane. *The First Astronomers: How Indigenous Elders Read the Stars.* Allen and Unwin, 2022.

Kepler, Johannes. *Somnium.* 1608.

Starkey, Natalie. *Catching Stardust: Comets, Asteroids and the Birth of the Solar System.* Bloomsbury Sigma, 2018.

Zubrin, Robert, and Christopher McKay. "Technological Requirements for Terraforming Mars." American Institute of Aeronautics and Astronautics, 2012.

PART THREE

Doyle, Arthur Conan. "The Horror of the Heights." *Strand Magazine* 46, no. 275 (1913).

Huygens, Christiaan. *Cosmotheoros: Or, Conjectures Concerning the Inhabitants of the Planets.* 1698.

———. *Treatise on Light.* 1690.

Maxwell, James Clerk. "A Dynamical Theory of the Electromagnetic Field." *Philosophical Transactions of the Royal Society,* 1865.

Swenson Jr., Loyd S. *Ethereal Aether: A History of the Michelson-Morley-Miller Aether-Drift Experiments, 1880–1930.* University of Texas Press Austin, 1972.

PART FOUR

Gott, J. Richard. *Time Travel in Einstein's Universe.* Houghton Mifflin, 2001.

Greene, Brian. *Light Falls: Space, Time, and an Obsession of Einstein* (audiobook). Audible Studios, 2016.

Hawking, Stephen. "Chronology Protection Conjecture." *Physical Review D* 46, no. 603 (1992).

Levin, Janna. *Black Hole Survival Guide.* Knopf, 2020.

———. *How the Universe Got Its Spots.* Princeton University Press, 2002.

Lossev, A., and I. D. Novikov. "The Jinn of the Time Machine: Nontrivial Self-Consistent Solutions." *Classical and Quantum Gravity* 9, no. 10 (1992).

Thorne, Kip S. *Black Holes and Time Warps: Einstein's Outrageous Legacy.* W. W. Norton & Co., 1994.

———. *The Science of* Interstellar. W. W. Norton & Co., 2014.

Wheeler, John Archibald. *Geons, Black Holes, and Quantum Foam.* W. W. Norton & Co., 1998.

ILLUSTRATIONS CREDITS

Cover, Nick Liefhebber; 3, A. Ghizzi Panizza/ESO; 6, NASA/JSC; 8, Corey Ford/Stocktrek Images/Science Source; 12–3, NASA, ESA, CSA, STScI; 14, NASA/Goddard Space Flight Center/Reto Stöckli; 16, Miguel Claro/Science Source; 20, Sergio Anelli/Electa/Mondadori Portfolio/ Getty Images; 25, NASA/JSC; 27, Steven Kazlowski/Nature Picture Library/Alamy Stock Photo; 32, Science & Society Picture Library/Getty Images; 35, krunja/Adobe Stock; 36, Photo © GraphicaArtis/Bridgeman Images; 38, Jay Nemeth/Red Bull Stratos; 41, Carlos Clarivan/Science Source; 44–5, NASA/JPL-Caltech; 50, Hans Strand/Folio Images/ Alamy Stock Photo; 55, NASA; 60, Michael Seeley; 62, NASA/MSFC; 64–5, NASA/Bob Nye; 66, Mark Thiessen/National Geographic Image Collection; 70, Mikkel Juul Jensen/Science Source; 75, Glenn Clovis; 78–9, Photo illustration by Marc Ward, with elements from NASA/ Shutterstock; 81, NASA/JPL-Caltech; 82, Dr. J. Durst/Science Source; 84, Science Source; 89, NASA/SDO; 90, Science & Society Picture Library/Getty Images; 94–5, NASA/Johns Hopkins University Applied Physics Laboratory/Carnegie Institution of Washington; 99, Everett Collection; 104, Eckhard Slawik/Science Source; 111, Christian Jegou/ Science Source; 114, Lynette Cook/Science Source; 119, Mark Garlick/ Science Source; 123, New York Public Library/Science Source; 126, Mark Garlick/Science Source; 129, Lowell Observatory Archives; 132–3, Photo illustration by Stockbym, with elements from NASA/Shutter-stock; 138, NASA/JPL-Caltech; 141, NASA/Johns Hopkins APL; 144, Enhanced Image by Gerald Eichstädt and Seán Doran (CC BY-NC-SA) based on images provided courtesy of NASA/JPL-Caltech/SwRI/MSSS; 150, NASA/JPL/Space Science Institute; 154, Ron Miller/Science Source; 158, NASA/Erich Karkoschka (Univ. Arizona); 161, NASA/Johns Hopkins University Applied Physics Laboratory/Southwest Research Institute; 164, ESA/Hubble & NASA, R. Sahai; 166, Everett Collection;

169, Mark Garlick/Science Source; 172, Equinox Graphics/Science Source; 175, © 2019-2022 CERN; 176, © Giancarlo Costa/Bridgeman Images; 181, © Fabrizio Carbone/EPFL; 183, George Karbus Photography/Cultura Creative RF/Alamy Stock Photo; 184, U.S. Naval History and Heritage Command, painting by Cliff Young; 189, Mark Garlick/Science Source; 193, NASA/Science Source; 194, Jose Antonio Peñas/Science Source; 198, NASA, ESA, CSA, STScI; 200–1, NASA, ESA, J. Hester and A. Loll (Arizona State University); 205, © Universal/courtesy Everett Collection; 206, Mark Garlick/Science Source; 212, artpartner-images/Getty Images; 215, ESO; 218, NASA/W. Stenzel; 220–1, NASA; 225, Design and illustration by Steve Burg; 226, Illustration by Sinelab; 230–1, ESA/Hubble & NASA; 232, Henning Dalhoff/Science Source; 234, Science: NASA, ESA, CSA, STScI. Image processing: Joseph DePasquale (STScI), Anton M. Koekemoer (STScI), Alyssa Pagan (STScI); 237, Mark Garlick/Science Source; 239, Christie's via Wikimedia Commons; 244, NASA, ESA, and the Hubble Heritage Team (STScI/AURA)—Hubble/Europe Collaboration. Acknowledgment: H. Bond (STScI and Penn State University); 247, Composite illustration by Tommaso Giannantonio. Credits: Earth: NASA/BlueEarth; Milky Way: ESO/S. Brunier; CMB: NASA/WMAP; 248, Pablo Carlos Budassi; 251, SPL/Science Source; 254–5, The SXS (Simulating eXtreme Spacetimes) Project; 256, Rendix Alextian/Adobe Stock; 257, Event Horizon Telescope Collaboration; 258, CBW/Alamy Stock Photo; 261, Autumn Blaze Creations; 264, Mark Garlick/Science Source; 267, Photo illustration by andrey_l, with elements from NASA/Shutterstock; 270, Henning Dalhoff/Science Source; 274, KTSDESIGN/Science Source; 279, Gregoire Cirade/Science Source; 280–1, Courtesy of Lucasfilm Ltd. — STAR WARS: A New Hope© & ™ Lucasfilm Ltd.; 282, Yuichiro Chino/Getty Images; 288, Stephen Hawking's Time Travellers Invitation— © Peter Dean; 289, Album/Alamy Stock Photo; 291, © Orion/courtesy Everett Collection; 293, Victor de Schwanberg/Science Source; 295, Courtesy of Marvel Studios © 2023 MARVEL; 297, GIROSCIENCE/Science Source; 300, Mary Evans/Ronald Grant/Everett Collection (10300793); 302–3, NASA, ESA, CSA, STScI, Webb ERO Production Team.

INDEX